Microbe
Hunters

Microbe Hunters

Paul de Kruif

Introduction by
F. Gonzalez-Crussi

A Harvest Book
Harcourt Brace & Company
San Diego   New York   London

To Rhea

Contents

PUBLISHER'S NOTE

Although certain words and phrases
contained herein may strike today's reader
as infelicitous, please bear in mind that
Microbe Hunters was first published in 1926
and is, of course, reflective of the author's
style as well as his time.

Introduction

F. Gonzalez-Crussi, M.D.

It does not happen very often that a book read in our youth should quicken, when reread after many years, the lively impressions that first moved us. Indeed, the opposite most commonly occurs—namely, that the volume fills us with a sort of puzzled disappointment, a melancholy feeling of defeated expectations. We are prompted to ask, "How could I have liked this stuff?" or "What did I see then that lies hidden from me now?," and for a while we appear to be at a loss for an answer. On those occasions, vanity generally finds an explanation (it almost always does): we have grown more sophisticated, wiser, with the accruing of experience, and our reading taste has become more difficult to satisfy. It also happens, although rarely, that a self-deprecating humility leads us to ask whether the years have made us worse. In other words, we wonder whether our present dislike or indifference results from the progressive hardening of our critical faculty into cynical faultfinding: a symptom, alas, of the withering of youthful high spirits.

It is thus no small praise to say of Paul de Kruif's *Microbe Hunters* that, well over half a century after it first saw the light

(1926), it manages to delight, and frequently to entrance, old and new readers—those who have kept a more or less blurry recollection of its pages from adolescence, as I had, and those for whom its vivid images and portrayals are a fresh experience. It continues to engage our hearts and minds today with an indescribable brand of affectionate sympathy. Survival of this magnitude is uncommon for any book, but especially for one aptly regarded as a work of "scientific popularization" and that deals with facts and personalities whose description has been reiterated ad nauseam. This appeal seems worthy of a closer look.

The writer's sincere identification with his subject is detectable throughout. De Kruif is truly elated at the discoveries, and despondent at the frustrations, of the men he follows. "How I wish I could take myself back, could bring you back to that innocent time when . . . How marvelous it would be to step into that simple Dutchman's [Leeuwenhoek's] shoes, to be inside his brain and body, to feel his excitement. . . ." Any less contagious or genuine enthusiasm would fail to do justice to the extraordinary epic with which the book opens, the unveiling of the microscopic world. For merely to say that the discovery of the microscope enlarged the scope of our vision sounds utterly pedestrian. It did much more than that: it replaced the world we lived in with a "plurality of worlds," each of them an abyss, a labyrinth, a universe replete with its own beauties and its own terrors. We used to see from the elephant down to the mite; thenceforth we had a world populated with tiny animalcules to whom the mite was elephant. Nothing is so solid that it lacks gaps or fractures, which to these tiny creatures are enormous cliffs and precipices; nothing is so uniform that it has no degree of fluidity. The seventeenth-century Dutch naturalist Swammerdam rhapsodized: "O God of miracles! How wonderful are thy works! . . . How well adapted the powers which thou hast so profusely bestowed upon all thy creatures!" But very soon the eulogy was tempered with the realization that the microscopic creatures can inflict upon humankind torments and agonies unsuspected, sufferings unheard of, pains and diseases yet unnamed. Hence the rhapsody must end on a somber note: Their marvelous organization notwithstanding, all living beings are subject to decay and destruction; "and, with all their perfections, scarce deserve to be considered shadows of the Divine Nature. It is therefore with the highest reason that [it has been said] all Nature is over-run and covered with a kind of leprosy. . . ."[1] It took a special breed of men to devise the means to protect us from the invisible, all-pervading, mortal threat. The lives of these men, their obsessions, their triumphs and defeats, form the rich tapestry of *Microbe Hunters*.

De Kruif uses a simple style that renders the book accessible to a wide public, including young readers. But we should not let ourselves be misled into confusing the simple, direct expression with a lowering of writing standards. There are no concessions to an ill-understood concept of "juvenile" literature. It would have been easier to narrate the chronological and circumstantial events: There are plenty of these to spin a good yarn. But our chronicler does not shun ideas, thoughts, and opinions. Ideas, even the most abstract, he approaches unassumingly and talks about unaffectedly. We sometimes lose sight of them because we encounter them in simple, almost homely attire: "A scientist, a really original investigator of nature, is like a writer, or a painter or a musician. He is part artist, part cool searcher. Spallanzani told himself stories. . . ." Simple as that.

Torrents of ink have been spent, from Aristotle on down, in disputing whether science and art act in concert or in opposition; whether the inventive dreamlike faculty of the artist is silenced or enhanced by the rational–critical bent of the scientist. Keats lamented that science unweaves the rainbow and makes a dull ordinariness out of what is solemn and awe-inspiring. Aldous Huxley sighed for a utopian future in which scientist and artist would walk hand in hand "into the ever

---

[1] Jan Swammerdam. *The Book of Nature*. Translated by Thomas Flloyd. London, 1758. Part 2, page 20.

expanding regions of the unknown." Philosophers, such as Karl Popper, have asserted that scientific thought is entirely accountable to reason and owes nothing to imagination. Scientists, such as the British Nobel laureate immunologist Peter Medawar (who as a literary writer was the equal of the best English authors in power and variety), have countered that scientific understanding always starts as an imaginative effort, a speculative leap that reconstructs what *might* be true—"a preconception which always, and necessarily, goes a little way (sometimes a long way) beyond anything which we have logical or factual authority to believe in."[2] In other words, the scientist begins by telling himself stories, just as Paul de Kruif tells us in *Microbe Hunters*, except that he says it without the solemnity of the above-quoted writers: "Great advances of science so often start from prejudice, on ideas got not from science, but straight out of the scientist's head, on notions that are only the opposite of the prevailing superstitious nonsense of the day."

Even though artist and scientist start from a common fount of creative imagination, their jobs soon place them on different, and in some ways diverging, tracks. Once the hypothesis is designed and experiments are under way, the scientist must stick to the information conveyed by the senses. Rigid standards of verification, systematic attempts at falsification, additional proofs and counterproofs, are deliberately enforced upon the scientist's constructs. The evidence must satisfy all of the observers, all of the time. For this job to be done well, courage, stubbornness, and more than average down-to-earthness are indispensable. The microbe hunters had to be as guarded against the general skepticism as against their own enthusiasm. Microscopy was fraught with the perils of both since its earliest beginnings. Undiscerning critics said of the microscopic world, when it was first revealed, that it consisted

---

[2] Peter B. Medawar. "Science and Literature." *The Hope of Progress: A Scientist Looks at Problems in Philosophy, Literature and Science.* New York: Doubleday, Anchor Books, 1973. 11–33.

of "vain and superfluous things" that could have no other use than "pomp and pleasure"; scholars doubted that there could be much truth learned from magnifying human vision and warned against error and misapprehension stemming from optical artifacts. In the opposite camp, charlatans claimed to possess lenses that revealed not just the details of a spider's leg, but the atoms of Epicurus, the subtle vapors that exhaled from the body, and the subtle impress left upon it by the influence of the stars.

The scientists portrayed in *Microbe Hunters* steer a difficult middle course between these extremes on their way to immortality. Their patient efforts are marvelously recapitulated. Their initial hypotheses and the rationale for their experiments, their gropings and failures, their sudden insights and joyous corroborations—all is laid out with the same simplicity. And the narrative's development at certain passages arouses our curiosity, increases our tension, and then satisfies it with a resolution that makes us say: "Why, of course! I see. Clever of Koch, or Ehrlich, or Pasteur, to have thought of that!"

No doubt sympathetic to his personages, the biographer nonetheless avoids hero-worship. The truly great scientists of the past have been dehumanized by dint of eulogies, institutionalized praise fashioning them into superhuman beings. Well, if we must have personality cults, and if contributions to the common weal sustain the candidatures, men such as Louis Pasteur clearly come ahead of most candidates. Here is a chemist who starts off by elucidating the molecular asymmetry of crystals; goes on to demonstrate that fermentation is due to living yeasts (and works out its mechanism, to the immense financial benefit of his country's wine industry); becomes a biologist, builds up the foundation of microbiological technique, and in the process deals a mortal blow to the foolish notion of spontaneous generation; improves the production of beer and eliminates the infectious hazards that plagued it; and rescues the silk industry from imminent disaster by diagnosing and preventing diseases of silkworms. And, wondrous to tell, he does all this *before* embarking on the feats for which he is most widely known: the discovery of the microbial cause of osteomyelitis, puerperal fever, pneumonia, fowl cholera, and sheep's anthrax; the invention of methods for attenuating the virulence of microbes and developing vaccines; and the triumph of his perseverance and intelligence against the dreadful terror of rabies.

De Kruif exposes all the grandeur of the microbe hunters' quiet toils. He is engrossing as he describes their intellectual exploits and the prodigality of their discoveries. He is inspiring in recalling the beneficent consequences of their pursuits. But he does not forget that the inquirers were human, and here lies much of the charm of this narrative. Spallanzani was a great scientist, but also a cunning, crafty wheeler-dealer; a man who avoided religious persecution by becoming a priest himself; a party declared innocent in legal suits brought against him, even though "it is not perfectly sure that he was not a little guilty." The distinguished Pasteur was not entirely above unseemly showmanship and petty jealousies, even at the zenith of his glory; and in his old age, rewarded with success and honor, strikes us as a pitiable invalid who must drag the paralyzed half of his body in order to receive the splendid guerdons. Walter Reed was a scholar and a sensitive man who freed humankind from yellow fever, one of the species's most terrible afflictions—but in laying out the groundwork for his extraordinary exploit, he performed experiments on human beings that cannot fail to evoke the sadistic researches carried out in Nazi concentration camps during the Second World War. In short, the microbe hunters were human. And, as such, they were very much a product of their time.

They could also be, as it is commonly said, ahead of their time. And it is their biographer's indisputable merit to have correctly apprehended some of their visionary insights. Consider de Kruif's treatment of Elie Metchnikoff. De Kruif is a popularizer whose colorful statements and familiar tone are apt to be distrusted by experts and academics. Thus, his characterization of Metchnikoff as a man who could be likened to "some hysterical personage out of Dostoevski's novels" and (more to the point here) a scientist who "in a manner of speaking founded immunology" tends to be dismissed as a highly romanticized misrepresentation. But for each of these statements no little factual evidence may be adduced. While it may be argued that no one person single-handedly built the theoretical basis of the complex scientific field of immunology, modern historical–philosophical scholarship points out that Metchnikoff's contributions were underestimated, even by members of the scientific community. For key biological concepts issue from his work, the full resonance of which is only now beginning to be appreciated.[3] Why do microbes, even in devastating epidemics, spare a few unlikely survivors? How do we become "tolerant" of foreign invaders or antigenic molecules? How does the organism discriminate between its own constitutive elements and "foreign" compounds? Metchnikoff's researches address the question of what is the biological "self." The very notion of selfhood, of what constitutes individual identity, lies ensconced in his views on immunity. And this combative man, saturated with the Darwinism of his day, saw immunity as a minidrama enacted in the interior of the human body from embryonic life to old age: an ever-renewing struggle between foreign invaders and defender cells, one of which, the phagocyte, became his lifelong obsession. Thus it may be said that if Metchnikoff is not *the* founder of immunology, he is the father of an important branch of this discipline, a body of knowledge that has retained a strange immediacy in our troubled times.

As the twentieth century draws to a close, new threats from the microbiologic world confront us, subtler and more formidable than those so painstakingly overcome. Some, such as

[3] See, especially, the work of Tauber and collaborators: Alfred I. Tauber. *The Immune Self: Theory or Metaphor.* New York: Cambridge University Press, 1994; A. I. Tauber and L. Chernyak. *Metchnikoff and the Origins of Immunology: From Metaphor to Theory.* New York: Oxford University Press, 1991; and Tauber and Chernyak. "The Birth of Immunology, part 2, Metchnikoff and His Critics." *Cellular Immunology* 121 (1989): 447–73.

the virus that causes AIDS, already raise apocalyptic visions of mass-scale havoc[4]; others, like the Ebola virus, only hint at new forms of widespread deadly torture. And as to former ills, their menace is not past. Foes that we believed subdued reappear again, as in the emergence of antibiotic-resistant bacterial strains. After a period of naive optimism, when we thought infectious diseases had been conquered, the stark realization dawns upon us that these remain the largest cause of death in the world, surpassing cardiovascular disease and cancer.[5] For it turns out that the efforts of the microbe hunters to secure us a definitive victory against the invisible foes earned us only a reprieve. The enemy, temporarily routed, rallies anew with uncanny new weapons drawn from the endless arsenal of evolutionary adaptation. No matter. A new generation of microbe hunters will thwart the invaders once more.

We must believe that human intelligence will serve us well this time, as it has served us in the past. We must believe that infectious diseases, whatever else they might be, are a *solvable* scientific and technological quandary. We must cling to the conviction that the new microbe hunters, heirs to those portrayed by de Kruif—here with unabashedly romantic brushstrokes, there from a naively teleologic viewpoint, but everywhere with engrossing style and flashes of keen insight —are going to be equal to the challenge. We must believe all this. For we have no other choice.

—Chicago, December 1995

---

[4] As of January 1993, 630,000 to 897,000 adults and adolescents were infected with AIDS in the United States. (See Philip S. Rosenberg. "Scope of the AIDS Epidemic in the United States." *Science* 270 (1995): 1372–75.)

[5] Barry R. Bloom and Christopher J. L. Murray. "Tuberculosis: Commentary on a Reemergent Killer." *Science* 257 (1992): 1055–64. See also: Harold C. Neu. "The Crisis in Antibiotic Resistance." *Science* 257 (1992): 1064–72; and Richard M. Krause. "The Origin of Plagues: Old and New." *Science* 257 (1992): 1073–78.

1

Leeuwenhoek

First of the Microbe Hunters

1

Two hundred and fifty years ago an obscure man named Leeuwenhoek looked for the first time into a mysterious new world
peopled with a thousand different kinds of tiny beings, some
ferocious and deadly, others friendly and useful, many of them
more important to mankind than any continent or archipelago.

Leeuwenhoek, unsung and scarce remembered, is now almost as unknown as his strange little animals and plants were
at the time he discovered them. This is the story of Leeuwenhoek, the first of the microbe hunters. It is the tale of the bold
and persistent and curious explorers and fighters of death who
came after him. It is the plain history of their tireless peerings
into this new fantastic world. They have tried to chart it, these
microbe hunters and death fighters. So trying they have
groped and fumbled and made mistakes and roused vain hopes.
Some of them who were too bold have died—done to death
by the immensely small assassins they were studying—and
these have passed to an obscure small glory.

To-day it is respectable to be a man of science. Those who
go by the name of scientist form an important element of the

population, their laboratories are in every city, their achievements are on the front pages of the newspapers, often before they are fully achieved. Almost any young university student can go in for research and by and by become a comfortable science professor at a tidy little salary in a cozy college. But take yourself back to Leeuwenhoek's day, two hundred and fifty years ago, and imagine yourself just through high school, getting ready to choose a career, wanting to know—

You have lately recovered from an attack of mumps, you ask your father what is the cause of mumps and he tells you a mumpish evil spirit has got into you. His theory may not impress you much, but you decide to make believe you believe him and not to wonder any more about what is mumps— because if you publicly don't believe him you are in for a beating and may even be turned out of the house. Your father is Authority.

That was the world three hundred years ago, when Leeuwenhoek was born. It had hardly begun to shake itself free from superstitions, it was barely beginning to blush for its ignorance. It was a world where science (which only means trying to find truth by careful observation and clear thinking) was just learning to toddle on vague and wobbly legs. It was a world where Servetus was burned to death for daring to cut up and examine the body of a dead man, where Galileo was shut up for life for daring to prove that the earth moved around the sun.

Antony Leeuwenhoek was born in 1632 amid the blue windmills and low streets and high canals of Delft, in Holland. His family were burghers of an intensely respectable kind and I say intensely respectable because they were basket-makers and brewers and brewers are respectable and highly honored in Holland. Leeuwenhoek's father died early and his mother sent him to school to learn to be a government official, but he left school at sixteen to be an apprentice in a dry-goods store in Amsterdam. That was his university. Think of a present-day scientist getting his training for experiment among bolts of gingham, listening to the tinkle of the bell on the cash drawer, being polite to an eternal succession of Dutch housewives who shopped with a penny-pinching dreadful exhaustiveness—but that was Leeuwenhoek's university, for six years!

At the age of twenty-one he left the dry-goods store, went back to Delft, married, set up a dry-goods store of his own there. For twenty years after that very little is known about him, except that he had two wives (in succession) and several children most of whom died, but there is no doubt that during this time he was appointed janitor of the city hall of Delft, and that he developed a most idiotic love for grinding lenses. He had heard that if you very carefully ground very little lenses out of clear glass, you would see things look much bigger than they appeared to the naked eye. . . . Little is known about him from twenty to forty, but there is no doubt that he passed in those days for an ignorant man. The only language he knew was Dutch—that was an obscure language despised by the cultured world as a tongue of fishermen and shopkeepers and diggers of ditches. Educated men talked Latin in those days, but Leeuwenhoek could not so much as read it and his only literature was the Dutch Bible. Just the same, you will see that his ignorance was a great help to him, for, cut off from all of the learned nonsense of his time, he had to trust to his own eyes, his own thoughts, his own judgment. And that was easy for him because there never was a more mulish man than this Antony Leeuwenhoek!

It would be great fun to look through a lens and see things bigger than your naked eye showed them to you! But *buy* lenses? Not Leeuwenhoek! There never was a more suspicious man. Buy lenses? He would make them himself! During these twenty years of his obscurity he went to spectacle-makers and got the rudiments of lens-grinding. He visited alchemists and apothecaries and put his nose into their secret ways of getting metals from ores, he began fumblingly to learn the craft of the gold- and silversmiths. He was a most pernickety man and was not satisfied with grinding lenses as good as those of the best lens-grinder in Holland, they had to be better than the best, and then he still fussed over them for long hours. Next he mounted these lenses in little oblongs of copper or silver or gold, which he had extracted himself, over hot fires, among strange smells and fumes. To-day searchers pay seventy-five dollars for a fine shining microscope, turn the screws, peer through it, make discoveries—without knowing anything about how it is built. But Leeuwenhoek—

Of course his neighbors thought he was a bit cracked but Leeuwenhoek went on burning and blistering his hands. Working forgetful of his family and regardless of his friends, he bent solitary to subtle tasks in still nights. The good neighbors sniggered, while that man found a way to make a tiny lens, less than one-eighth of an inch across, so symmetrical, so perfect, that it showed little things to him with a fantastic clear enormousness. Yes, he was a very uncultured man, but he alone of all men in Holland knew how to make those lenses, and he said of those neighbors: "We must forgive them, seeing that they know no better."

Now this self-satisfied dry-goods dealer began to turn his lenses onto everything he could get hold of. He looked through them at the muscle fibers of a whale and the scales of his own skin. He went to the butcher shop and begged or bought ox-eyes and was amazed at how prettily the crystalline lens of the eye of the ox is put together. He peered for hours at the build of the hairs of a sheep, of a beaver, of an elk, that were transformed from their fineness into great rough logs under his bit of glass. He delicately dissected the head of a fly; he stuck its brain on the fine needle of his microscope—how he admired the clear details of the marvelous big brain of that fly! He examined the cross-sections of the wood of a dozen different trees and squinted at the seeds of plants. He grunted "Impossible!" when he first spied the outlandish large perfection of the sting of a flea and the legs of a louse. That man Leeuwenhoek was like a puppy who sniffs—with a totally impolite disregard of discrimination—at every object of the world around him!

2

There never was a less sure man than Leeuwenhoek. He looked at this bee's sting or that louse's leg again and again and again. He left his specimens sticking on the point of his strange microscope for months—in order to look at other things he made more microscopes till he had hundreds of them!—then he came back to those first specimens to correct his first mistakes. He never set down a word about anything he peeped at, he never made a drawing until hundreds of peeps showed him that, under given conditions, he would always see exactly the same thing. And then he was not sure! He said:

"People who look for the first time through a microscope say now I see this and then I see that—and even a skilled observer can be fooled. On these observations I have spent more time than many will believe, but I have done them with joy, and I have taken no notice of those who have said why take so much trouble and what good is it?—but I do not write for such people but only for the philosophical!" He worked for twenty years that way, without an audience.

But at this time, in the middle of the seventeenth century, great things were astir in the world. Here and there in France and England and Italy rare men were thumbing their noses at almost everything that passed for knowledge. "We will no longer take Aristotle's say-so, nor the Pope's say-so," said these rebels. "We will trust only the perpetually repeated observations of our own eyes and the careful weighings of our scales; we will listen to the answers experiments give us and no other answers!" So in England a few of these revolutionists started a society called The Invisible College, it had to be invisible because that man Cromwell might have hung them for plotters and heretics if he had heard of the strange questions they were trying to settle. What experiments those solemn searchers made! Put a spider in a circle made of the powder of a unicorn's horn and that spider can't crawl out—so said the wisdom of that day. But these Invisible Collegians?

One of them brought what was supposed to be powdered unicorn's horn and another came carrying a little spider in a bottle. The college crowded around under the light of high candles. Silence, then the hushed experiment, and here is their report of it:

"A circle was made with the powder of unicorn's horn and a spider set in the middle of it, but it immediately ran out."

Crude, you exclaim. Of course! But remember that one of the members of this college was Robert Boyle, founder of the science of chemistry, and another was Isaac Newton. Such was the Invisible College, and presently, when Charles II came to the throne, it rose from its depths as a sort of blind-pig scientific society to the dignity of the name of the Royal Society of England. And they were Antony Leeuwenhoek's first audience! There was one man in Delft who did not laugh at Antony Leeuwenhoek, and that was Regnier de Graaf, whom the Lords and Gentlemen of the Royal Society had made a corresponding member because he had written them of interesting things he had found in the human ovary. Already Leeuwenhoek was rather surly and suspected everybody, but he let de Graaf peep through those magic eyes of his, those little lenses whose equal did not exist in Europe or England or the whole world for that matter. What de Graaf saw through those microscopes made him ashamed of his own fame and he hurried to write to the Royal Society:

"Get Antony Leeuwenhoek to write you telling of his discoveries."

And Leeuwenhoek answered the request of the Royal Society with all the confidence of an ignorant man who fails to realize the profound wisdom of the philosophers he addresses. It was a long letter, it rambled over every subject under the sun, it was written with a comical artlessness in the conversational Dutch that was the only language he knew. The title of that letter was: "A Specimen of some Observations made by a Microscope contrived by Mr. Leeuwenhoek, concerning Mould upon the Skin, Flesh, etc.; the Sting of a Bee, etc."

The Royal Society was amazed, the sophisticated and learned gentlemen were amused—but principally the Royal Society was astounded by the marvelous things Leeuwenhoek told them he could see through his new lenses. The Secretary of the Royal Society thanked Leeuwenhoek and told him he hoped his first communication would be followed by others. It was, by hundreds of others over a period of fifty years. They were talkative letters full of salty remarks about his ignorant neighbors, of exposures of charlatans and of skilled explodings of superstitions, of chatter about his personal health—but sandwiched between paragraphs and pages of this homely stuff, in almost every letter, those Lords and Gentlemen of the Royal Society had the honor of reading immortal and gloriously accurate descriptions of the discoveries made by the magic eye of that janitor and shopkeeper. What discoveries!

When you look back at them, many of the fundamental discoveries of science seem so simple, too absurdly simple. How was it men groped and fumbled for so many thousands of years without seeing things that lay right under their noses? So with microbes. Now all the world has seen them cavorting on movie screens, many people of little learning have peeped at them swimming about under lenses of microscopes, the greenest medical student is able to show you the germs of I don't know how many diseases—what was so hard about seeing microbes for the first time?

But let us drop our sneers to remember that when Leeuwenhoek was born there were no microscopes but only crude hand-lenses that would hardly make a ten-cent piece look as large as a quarter. Through these—without his incessant grinding of his own marvelous lenses—that Dutchman might have looked till he grew old without discovering any creature smaller than a cheese-mite. You have read that he made better and better lenses with the fanatical persistence of a lunatic; that he examined everything, the most intimate things and the most shocking things, with the silly curiosity of a puppy. Yes, and all this squinting at bee-stings and mustache hairs and what-not was needful to prepare him for that sudden day when he looked through his toy of a gold-mounted lens at a fraction of a small drop of clear rain water to discover—

What he saw that day starts this history. Leeuwenhoek was a maniac observer, and who but such a strange man would have thought to turn his lens on clear, pure water, just come down from the sky? What could there be in water but just—water? You can imagine his daughter Maria—she was nineteen and she took such care of her slightly insane father!—watching him take a little tube of glass, heat it red-hot in a flame, draw it out to the thinness of a hair. . . . Maria was devoted to her father—let any of those stupid neighbors dare to snigger at him!—but what in the world was he up to now, with that hair-fine glass pipe?

You can see her watch that absent-minded wide-eyed man break the tube into little pieces, go out into the garden to bend over an earthen pot kept there to measure the fall of the rain. He bends over that pot. He goes back into his study. He sticks the little glass pipe onto the needle of his microscope. . . .

What can that dear silly father be up to?

He squints through his lens. He mutters guttural words under his breath. . . .

Then suddenly the excited voice of Leeuwenhoek: "Come here! Hurry! There are little animals in this rain water. . . . They swim! They play around! They are a thousand times smaller than any creatures we can see with our eyes alone. . . . Look! See what I have discovered!"

Leeuwenhoek's day of days had come. Alexander had gone to India and discovered huge elephants that no Greek had ever seen before—but those elephants were as commonplace to Hindus as horses were to Alexander. Cæsar had gone to England and come upon savages that opened his eyes with wonder—but these Britons were as ordinary to each other as Roman centurions were to Cæsar. Balboa? What were his proud feelings as he looked for the first time at the Pacific? Just the same that Ocean was as ordinary to a Central American Indian as the Mediterranean was to Balboa. But Leeuwenhoek? This janitor of Delft had stolen upon and peeped into a fantastic sub-visible world of little things, creatures that had lived, had bred, had battled, had died, completely hidden from and unknown to all men from the beginning of time. Beasts these were of a kind that ravaged and annihilated whole races of men ten million times larger than they were themselves. Beings these were, more terrible than fire-spitting dragons or hydra-headed monsters. They were silent assassins that murdered babes in warm cradles and kings in sheltered places. It was this invisible, insignificant, but implacable—and sometimes friendly—world that Leeuwenhoek had looked into for the first time of all men of all countries.

This was Leeuwenhoek's day of days. . . .

3

That man was so unashamed of his admirations and his surprises at a nature full of startling events and impossible things. How I wish I could take myself back, could bring you back, to that innocent time when men were just beginning to disbelieve in miracles and only starting to find still more miraculous facts. How marvelous it would be to step into that simple Dutchman's shoes, to be inside his brain and body, to feel his excitement—it is almost nausea!—at his first peep at those cavorting "wretched beasties."

That was what he called them, and, as I have told you, this Leeuwenhoek was an unsure man. Those animals were too tremendously small to be true, they were too strange to be true. So he looked again, till his hands were cramped with holding his microscope and his eyes full of that smarting water that comes from too-long looking. But he was right! Here they were again, not one kind of little creature, but here was another, larger than the first, "moving about very nimbly because they were furnished with divers incredibly thin feet." Wait! Here is a third kind—and a fourth, so tiny I can't make out his shape. But he is alive! He goes about, dashing over great distances in this world of his water-drop in the little tube. . . . What nimble creatures!

"They stop, they stand still as 'twere upon a point, and then turn themselves round with that swiftness, as we see a top turn round, the circumference they make being no bigger than that of a fine grain of sand." So wrote Leeuwenhoek.

For all this seemingly impractical sniffing about, Leeuwenhoek was a hard-headed man. He hardly ever spun theories, he was a fiend for measuring things. Only how could you make a measuring stick for anything so small as these little beasts? He wrinkled his low forehead: "How large really is this last and smallest of the little beasts?" He poked about in the cobwebbed corners of his memory among the thousand other things he had studied with you can't imagine what thoroughness; he made calculations: "This last kind of animal is a thousand times smaller than the eye of a large louse!" That was an accurate man. For we know now that the eye of one full-grown louse is no larger nor smaller than the eyes of ten thousand of his brother and sister lice.

But where did these outlandish little inhabitants of the rain water come from? Had they come down from the sky? Had they crawled invisibly over the side of the pot from the ground? Or had they been created out of nothing by a God full of whims? Leeuwenhoek believed in God as piously as any Seventeenth Century Dutchman. He always referred to God as the Maker of the Great All. He not only believed in God but he admired him intensely—what a Being to know how to fashion bees' wings so prettily! But then Leeuwenhoek was a materialist too. His good sense told him that life comes from life. His simple belief told him that God had invented all living things in six days, and, having set the machinery going, sat back to reward good observers and punish guessers and bluffers. He stopped speculating about improbable gentle rains of little animals from heaven. Certainly God couldn't brew those animals in the rain water pot out of nothing! But wait . . . Maybe? Well, there was only one way to find out where they came from. "I will experiment!" he muttered.

He washed out a wine glass very clean, he dried it, he held it under the spout of his eaves-trough, he took a wee drop in one of his hair-fine tubes. Under his lens it went. . . . Yes! They were there, a few of those beasts, swimming about. . . . "They are present even in very fresh rain water!" But then, that really proved nothing, they might live in the eaves-trough and be washed down by the water. . . .

Then he took a big porcelain dish, "glazed blue within," he washed it clean, out into the rain he went with it and put it on top of a big box so that the falling raindrops would splash no mud into the dish. The first water he threw out to clean it still more thoroughly. Then intently he collected the next bit in one of his slender pipes, into his study he went with it. . . .

"I have proved it! This water has not a single little creature in it! They do not come down from the sky!"

But he kept that water; hour after hour, day after day he squinted at it—and on the fourth day he saw those wee beasts beginning to appear in the water along with bits of dust and little flecks of thread and lint. That was a man from Missouri! Imagine a world of men who would submit all of their cock-sure judgments to the ordeal of the common-sense experiments of a Leeuwenhoek!

Did he write to the Royal Society to tell them of this entirely unsuspected world of life he had discovered? Not yet! He was a slow man. He turned his lens onto all kinds of water, water kept in the close air of his study, water in a pot kept on the high roof of his house, water from the not-too-clean canals of Delft and water from the deep cold well in his garden. Everywhere he found those beasts. He gaped at their enormous littleness, he found many thousands of them did not equal a grain of sand in bigness, he compared them to a cheese-mite and they were to this filthy little creature as a bee is to a horse. He was never tired with watching them "swim about among one another gently like a swarm of mosquitoes in the air. . . ."

Of course this man was a groper. He was a groper and a

stumbler as all men are gropers, devoid of prescience, and stumblers, finding what they never set out to find. His new beasties were marvelous but they were not enough for him, he was always poking into everything, trying to see more closely, trying to find reasons. Why is the sharp taste of pepper? That was what he asked himself one day, and he guessed: "There must be little points on the particles of pepper and these points jab the tongue when you eat pepper. . . ."

But are there such little points?

He fussed with dry pepper. He sneezed. He sweat, but he couldn't get the grains of pepper small enough to put under his lens. So, to soften it, he put it to soak for several weeks in water. Then with fine needles he pried the almost invisible specks of the pepper apart, and sucked them up in a little drop of water into one of his hair-fine glass tubes. He looked—

Here was something to make even this determined man scatter-brained. He forgot about possible small sharp points on the pepper. With the interest of an intent little boy he watched the antics of "an incredible number of little animals, of various sorts, which move very prettily, which tumble about and sidewise, this way and that!"

So it was Leeuwenhoek stumbled on a magnificent way to grow his new little animals.

And now to write all this to the great men off there in London! Artlessly he described his own astonishment to them. Long page after page in a superbly neat handwriting with little common words he told them that you could put a million of these little animals into a coarse grain of sand and that one drop of his pepper-water, where they grew and multiplied so well, held more than two-million seven-hundred-thousand of them. . . .

This letter was translated into English. It was read before the learned skeptics—who no longer believed in the magic virtues of unicorn's horns—and it bowled the learned body over! What! The Dutchman said he had discovered beasts so small that you could put as many of them into one little drop of water as there were people in his native country? Nonsense!

The cheese-mite was absolutely and without doubt the smallest creature God had created.

But a few of the members did not scoff. This Leeuwenhoek was a confoundedly accurate man: everything he had ever written to them they had found to be true. . . . So a letter went back to the scientific janitor, begging him to write them in detail the way he had made his microscope, and his method of observing.

That upset Leeuwenhoek. It didn't matter that these stupid oafs of Delft laughed at him—but the Royal Society? He had thought *they* were philosophers! Should he write them details, or should he from now on keep everything he did to himself? "Great God," you can imagine him muttering, "these ways I have of uncovering mysterious things, how I have worked and sweat to learn to do them, what jeering from how many fools haven't I endured to perfect my microscopes and my ways of looking! . . ."

But creators must have audiences. He knew that these doubters of the Royal Society should have sweat just as hard to disprove the existence of his little animals as he himself had toiled to discover them. He was hurt, but—creators must have an audience. So he replied to them in a long letter assuring them he never told anything too big. He explained his calculations (and modern microbe hunters with all of their apparatus make only slightly more accurate ones!) he wrote these calculations out, divisions, multiplications, additions, until his letter looked like a child's exercise in arithmetic. He finished by saying that many people of Delft had seen—with applause!—these strange new animals under his lens. He would send them affidavits from prominent citizens of Delft —two men of God, one notary public, and eight other persons worthy to be believed. But he wouldn't tell them how he made his microscopes.

That was a suspicious man! He held his little machines up for people to look through, but let them so much as touch the microscope to help themselves to see better and he might order them out of his house. . . . He was like a child anxious and proud to show a large red apple to his playmates but loth to let them touch it for fear they might take a bite out of it.

So the Royal Society commissioned Robert Hooke and Nehemiah Grew to build the very best microscopes, and brew pepper water from the finest quality of black pepper. And, on the 15th of November, 1677, Hooke came carrying his microscope to the meeting—agog—for Antony Leeuwenhoek had not lied. Here they were, those enchanted beasts! The members rose from their seats and crowded round the microscope. They peered, they exclaimed: this man must be a wizard observer! That was a proud day for Leeuwenhoek. And a little later the Royal Society made him a Fellow, sending him a gorgeous diploma of membership in a silver case with the coat of arms of the society on the cover. "I will serve you faithfully during the rest of my life," he wrote them. And he was as good as his word, for he mailed them those conversational mixtures of gossip and science till he died at the age of ninety. But send them a microscope? Very sorry, but that was impossible to do, while he lived. The Royal Society went so far as to dispatch Doctor Molyneux to make a report on this janitor-discoverer of the invisible. Molyneux offered Leeuwenhoek a fine price for one of his microscopes—surely he could spare one?—for there were hundreds of them in cabinets that lined his study. But no! Was there anything the gentleman of the Royal Society would like to see? Here were some most curious little unborn oysters in a bottle, here were divers very nimble little animals, and that Dutchman held up his lenses for the Englishman to peep through, watching all the while out of the corner of his eye to see that the undoubtedly most honest visitor didn't touch anything—or filch anything. . . .

"But your instruments are marvelous!" cried Molyneux. "A thousand times more clear they show things than any lens we have in England!"

"How I wish, Sir," said Leeuwenhoek, "that I could show you my best lens, with my special way of observing, but I keep that only for myself and do not show it to any one—not even to my own family."

4

Those little animals were everywhere! He told the Royal Society of finding swarms of those sub-visible beings in his mouth—of all places: "Although I am now fifty years old," he wrote, "I have uncommonly well-preserved teeth, because it is my custom every morning to rub my teeth very hard with salt, and after cleaning my large teeth with a quill, to rub them vigorously with a cloth. . . ." But there still were little bits of white stuff between his teeth, when he looked at them with a magnifying mirror. . . .

What was this white stuff made of?

From his teeth he scraped a bit of this stuff, mixed it with pure rain water, stuck it in a little tube on to the needle of his microscope, closed the door of his study—

What was this that rose from the gray dimness of his lens into clear distinctness as he brought the tube into the focus?

Here was an unbelievably tiny creature, leaping about in the water of the tube "like the fish called a pike." There was a second kind that swam forward a little way, then whirled about suddenly, then tumbled over itself in pretty somersaults. There were some beings that moved sluggishly and looked like wee bent sticks, nothing more, but that Dutchman squinted at them till his eyes were red-rimmed—and they moved, they were alive, no doubt of it! There was a menagerie in his mouth! There were creatures shaped like flexible rods that went to and fro with the stately carriage of bishops in procession, there were spirals that whirled through the water like violently animated corkscrews. . . .

Everybody he could get hold of—as well as himself—was an experimental animal for that curious man. Tired from his long peering at the little beasts in his own mouth, he went for a walk under the tall trees that dropped their yellow leaves on the brown mirrors of the canals; it was hard work, this play of his, he must rest! But he met an old man, a most interesting old man: "I was talking to this old man," wrote Leeuwenhoek to the Royal Society, "an old man who led a very sober life, who never used brandy nor tobacco and very seldom wine, and my eye chanced to fall on his teeth which were badly grown over and that made me ask him when he had last cleaned his mouth. I got for answer that he had never cleaned his teeth in his whole life. . . ."

Away went all thought of his aching eyes. What a zoo of wee animals must be in this old fellow's mouth. He dragged the dirty but virtuous victim of his curiosity into his study— of course there were millions of wee beasties in that mouth, but what he wanted particularly to tell the Royal Society was this: that this old man's mouth was host to a new kind of creature, that slid along among the others, bending its body in graceful bows like a snake—the water in the narrow tube seemed to be alive with those little fellows!

You may wonder that Leeuwenhoek nowhere in any of those hundreds of letters makes any mention of the harm these mysterious new little animals might do to men. He had come upon them in drinking water, spied upon them in the mouth; as the years went by he discovered them in the intestines of frogs and horses, and even in his own discharges; in swarms he found them on those rare occasions when, as he says, "he was troubled with a looseness." But not for a moment did he guess that his trouble was caused by those little beasts, and from his unimaginativeness and his carefulness not to jump to conclusions modern microbe hunters—if they only had time to study his writings—could learn a great deal. For, during the last fifty years, literally thousands of microbes have been described as the authors of hundreds of diseases, when, in the majority of cases those germs have only been chance residents in the body at the time it became diseased. Leeuwenhoek was cautious about calling anything the *cause* of anything else. He had a sound instinct about the infinite complicatedness of everything—that told him the danger of trying to pick out one cause from the tangled maze of causes which control life. . . .

The years went by. He tended his little dry-goods store, he saw to it the city hall of Delft was properly swept out, he grew more and more crusty and suspicious, he looked longer and longer hours through his hundreds of microscopes, he made a hundred amazing discoveries. In the tail of a little fish stuck head first into a glass tube he saw for the first time of all men the capillary blood vessels through which blood goes from the arteries to the veins—so he completed the Englishman Harvey's discovery of the circulation of the blood. The most sacred and improper and romantic things in life were only material for the probing, tireless eyes of his lenses. Leeuwenhoek discovered the human sperm, and the cold-blooded science of his searching would have been shocking, if he had not been such a completely innocent man! The years went by and all Europe knew about him. Peter the Great of Russia came to pay his respects to him, and the Queen of England journeyed to Delft only to look at the wonders to be seen through the lenses of his microscopes. He exploded countless superstitions for the Royal Society, and aside from Isaac Newton and Robert Boyle he was the most famous of their members. But did these honors turn his head? They couldn't turn his head because he had from the first a sufficiently high opinion of himself! His arrogance was limitless—but it was equaled by his humility when he thought of that misty unknown that he knew surrounded himself and all men. He admired the Dutch God but his real god was truth:

"My determination is not to remain stubbornly with my ideas but I'll leave them and go over to others as soon as I am shown plausible reasons which I can grasp. This is the more true since I have no other purpose than to place truth before my eyes so far as it is in my power to embrace it; and to use the little talent I have received to draw the world away from its old heathenish superstitions and to go over to the truth and to stick to it."

He was an amazingly healthy man, and at the age of eighty his hand hardly trembled as he held up his microscope for visitors to peep at his little animals or to exclaim at the unborn oysters. But he was fond of drinking in the evenings—as what Dutchman is not?—and his only ill seems to have been a certain seediness in the morning after such wassail. He detested physicians—how could they know about the ills of the body when they didn't know one thousandth of what he did about the build of the body? So Leeuwenhoek had his own theories—and sufficiently foolish they were—about the cause of this seediness. He knew that his blood was full of little globules—he had been the first of all men to see them. He knew those globules had to go through very tiny capillaries to get from his arteries to his veins—hadn't he been the man to discover those wee vessels in a fish tail? Well, after those hilarious nights of his, his blood got too thick to run properly from the arteries to the veins! So he would thin it! So he wrote to the Royal Society:

"When I have supped too heavily of an evening, I drink in the morning a large number of cups of coffee, and that as hot as I can drink it, so that the sweat breaks out on me, and if by so doing I can't restore my body, a whole apothecary's shop couldn't do much, and that is the only thing I have done for years when I have felt a fever."

That hot coffee drinking led him to another curious fact about the little animals. Everything he did led him to pry up some new fact of nature, for he lived wrapped in those tiny dramas that went on under his lenses just as a child listens open-mouthed with saucer eyes to the myths of Mother Goose. . . . He never tired of reading the same story of nature, there were always new angles to be found in it, the pages of his book of nature were thumbed and dog-eared by his insatiable interest. Years after his discovery of the microbes in his mouth one morning in the midst of his sweating from his vast

curative coffee drinkings he looked once more at the stuff between his teeth—

What was this? There was not a single little animal to be
found. Or there were no living animals rather, for he thought
he could make out the bodies of myriads of dead ones—and
maybe one or two that moved feebly, as if they were sick.
"Blessed Saints!" he growled: "I hope some great Lord of the
Royal Society doesn't try to find those creatures in his mouth,
and fail, and then deny my observations. . . ."

But look here! He had been drinking coffee, so hot it had
blistered his lips, almost. He had looked for the little animals
in the white stuff from between his front teeth. It was just
after the coffee he had looked there— Well?

With the help of a magnifying mirror he went at his back
teeth. Presto! "With great surprise I saw an incredibly large
number of little animals, and in such an unbelievable quantity
of the aforementioned stuff, that it is not to be conceived of
by those who have not seen it with their own eyes." Then he
made delicate experiment in tubes, heating the water with its
tiny population to a temperature a little warmer than that of
a hot bath. In a moment the creatures stopped their agile run-
nings to and fro. He cooled the water. They did not come
back to life—so! It was that hot coffee that had killed the
beasties in his front teeth!

With what delight he watched them once more! But he was
bothered, he was troubled, for he couldn't make out the heads
or tails of any of his little animals. After wiggling forward in
one direction they stopped, they reversed themselves and
swam backward just as swiftly without having turned around.
But they *must* have heads and tails! They must have livers and
brains and blood vessels as well! His thoughts floated back to
his work of forty years before, when he had found that under
his powerful lenses fleas and cheese-mites, so crude and simple
to the naked eye, had become as complicated and as perfect
as human beings. But try as he would, with the best lenses
he had, and those little animals in his mouth were just plain
sticks or spheres or corkscrews. So he contented himself by

calculating, for the Royal Society, what the diameter of the invisible blood vessels of his microbes must be—but mind you, he never for a moment hinted that he had seen such blood vessels; it only amused him to stagger his patrons by speculations of their unthinkable smallness.

If Antony Leeuwenhoek failed to see the germs that cause human disease, if he had too little imagination to predict the rôle of assassin for his wretched creatures, he did show that sub-visible beasts could devour and kill living beings much larger than they were themselves. He was fussing with mussels, shellfish that he dredged up out of the canals of Delft. He found thousands of them unborn inside their mothers. He tried to make these young ones develop outside their mothers in a glass of canal water. "I wonder," he muttered, "why our canals are not choked with mussels, when the mothers have each one so many young ones inside them!" Day after day he poked about in his glass of water with its slimy mass of embryos, he turned his lens on to them to see if they were growing—but what was this? Astounded he watched the fishy stuff disappear from between their shells—it was being gobbled up by thousands of tiny microbes that were attacking the mussels greedily. . . .

"Life lives on life—it is cruel, but it is God's will," he pondered. "And it is for our good, of course, because if there weren't little animals to eat up the young mussels, our canals would be choked by those shellfish, for each mother has more than a thousand young ones at a time!" So Antony Leeuwenhoek accepted everything and praised everything, and in this he was a child of his time, for in his century searchers had not yet, like Pasteur who came after them, begun to challenge God, to shake their fists at the meaningless cruelties of nature toward mankind, her children. . . .

He passed eighty, and his teeth came loose as they had to even in his strong body; he didn't complain at the inexorable arrival of the winter of his life, but he jerked out that old tooth and turned his lens onto the little creatures he found within that hollow root—why shouldn't he study them once more?

There might be some little detail he had missed those hundred other times! Friends came to him at eighty-five and told him to take it easy and leave his studies. He wrinkled his brow and opened wide his still bright eyes: "The fruits that ripen in autumn last the longest!" he told them—he called eighty-five the autumn of his life!

Leeuwenhoek was a showman. He was very pleased to hear the ohs and ahs of people—they must be philosophical people and lovers of science, mind you!—whom he let peep into his sub-visible world or to whom he wrote his disjointed marvelous letters of description. But he was no teacher. "I've never taught one," he wrote to the famous philosopher Leibniz, "because if I taught one, I'd have to teach others. . . . I would give myself over to a slavery, whereas I want to stay a free man."

"But the art of grinding fine lenses and making observations of these new creatures will disappear from the earth, if you don't teach young men," answered Leibniz.

"The professors and students of the University of Leyden were long ago dazzled by my discoveries, they hired three lens grinders to come to teach the students, but what came of it?" wrote that independent Dutchman.

"Nothing, so far as I can judge, for almost all of the courses they teach there are for the purpose of getting money through knowledge or for gaining the respect of the world by showing people how learned you are, and these things have nothing to do with discovering the things that are buried from our eyes. I am convinced that of a thousand people not one is capable of carrying out such studies, because endless time is needed and much money is spilled and because a man has always to be busy with his thoughts if anything is to be accomplished. . . ."

That was the first of the microbe hunters. In 1723, when he was ninety-one years old and on his deathbed, he sent for his friend Hoogvliet. He could not lift his hand. His once glowing eyes were rheumy and their lids were beginning to stick fast with the cement of death. He mumbled:

"Hoogvliet, my friend, be so good as to have those two

letters on the table translated into Latin. . . . Send them to London to the Royal Society. . . ."

So he kept his promise made fifty years before, and Hoog-vliet wrote, along with those last letters: "I send you, learned sirs, this last gift of my dying friend, hoping that his final word will be agreeable to you."

So he passed, this first of the microbe hunters. You will read of Spallanzani, who was much more brilliant, of Pasteur who had a thousand times his imagination, of Robert Koch who did much more immediate apparent good in lifting the torments that microbes bring to men—these and all the others have much more fame to-day. But not one of them has been so completely honest, so appallingly accurate as this Dutch janitor, and all of them could take lessons from his splendid common sense.

2

Spallanzani

Microbes Must Have Parents!

1

"Leeuwenhoek is dead, it is too bad, it is a loss that cannot be made good. Who now will carry on the study of the little animals?" asked the learned men of the Royal Society in England, asked Réaumur and the brilliant Academy in Paris. Their question did not wait long for an answer, for the janitor of Delft had hardly closed his eyes in 1723 for the long sleep that he had earned so well, when another microbe hunter was born, in 1729 a thousand miles away in Scandiano in northern Italy. This follower of Leeuwenhoek was Lazzaro Spallanzani, a strange boy who lisped verses while he fashioned mudpies; who forgot mudpies to do fumbling childish and cruel experiments with beetles and bugs and flies and worms. Instead of pestering his parents with questions, he examined living things in nature, by pulling legs and wings off them, by trying to stick them back on again. He must find out how things worked; he didn't care so very much what they looked like.

Like Leeuwenhoek, the young Italian had to fight to become a microbe hunter against the wishes of his family. His father was a lawyer and did his best to get Lazzaro interested in long sheets of legal foolscap—but the youngster sneaked away and skipped flat stones over the surface of the water, and wondered why the stones skipped and didn't sink.

In the evenings he was made to sit down before dull lessons, but when his father's back was turned he looked out of the window at the stars that gleamed in the velvet black Italian sky, and next morning lectured about them to his playmates until they called him "The Astrologer."

On holidays he pushed his burly body through the woods near Scandiano, and came wide-eyed upon foaming natural fountains. These made him stop his romping, and caused him to go home sunk in unboyish thought. What caused these fountains? His folks and the priest had told him they had sprung in olden times from the tears of sad, deserted, beautiful girls who were lost in the woods. . . .

Lazzaro was a dutiful son—and a politician of a son—so he didn't argue with his father or the priest. But to himself he said "bunk" to their explanation, and made up his mind to find out, some day, the real why and wherefore of fountains.

Young Spallanzani was just as determined as Leeuwenhoek had been to find out the hidden things of nature, but he set about getting to be a scientist in an entirely different way. He pondered: "My father insists that I study law, does he?" He kept up the pretense of being interested in legal documents—but in every spare moment he boned away at mathematics and Greek and French and Logic—and during his vacations watched skipping stones and fountains, and dreamed about understanding the violent fireworks of volcanoes. Then craftily he went to the noted scientist, Vallisnieri, and told this great man what he knew. "But you were born for a scientist," said Vallisnieri, "you waste time foolishly, studying lawbooks."

"Ah, master, but my father insists."

Indignantly Vallisnieri went to Spallanzani senior and scolded him for throwing away Lazzaro's talents on the merely useful study of law. "Your boy," he said, "is going to be a searcher, he will honor Scandiano, and make it famous—he is like Galileo!"

And the shrewd young Spallanzani went to the University at Reggio, with his father's blessing, to take up the career of scientist.

At this time it was much more respectable and safe to be a scientist than it had been when Leeuwenhoek began his first grinding of lenses. The Grand Inquisition was beginning to pull in its horns. It preferred jerking out the tongues of obscure alleged criminals and burning the bodies of unknown heretics, to persecuting Servetuses and Galileos. The Invisible College no longer met in cellars or darkened rooms, and learned societies all over were now given the generous support of parliaments and kings. It was not only beginning to be permitted to question superstitions, it was becoming fashionable to do it. The thrill and dignity of real research into nature began to elbow its way into secluded studies of philosophers. Voltaire retired for years into the wilds of rural France to master the great discoveries of Newton, and then to popularize them in his country. Science even penetrated into brilliant and witty and immoral drawing-rooms, and society leaders like Madame de Pompadour bent their heads over the forbidden Encyclopedia—to try to understand the art and science of the making of rouge and silk stockings.

Along with this excited interest in everything from mechanics of the stars to the caperings of little animals, the people of Spallanzani's glittering century began to show an open contempt for religion and dogmas, even the most sacred ones. A hundred years before men had risked their skins to laugh at the preposterous and impossible animals that Aristotle had gravely put into his books on biology. But now, they could openly snicker at the mention of his name and whisper: "Because he's Aristotle it implies that he must be believed e'en though he lies." Still there was plenty of ignorance in the world, and much pseudo-science—even in the Royal Societies and Academies. And Spallanzani, freed from the horror of an endless future of legal wranglings, threw himself with vigor into getting all kinds of knowledge, into testing all kinds of theories, into disrespecting all kinds of authorities no matter how famous, into association with every kind of person, from fat bishops, officials, and professors to outlandish actors and minstrels.

He was the very opposite of Leeuwenhoek, who so patiently had ground lenses, and looked at everything for twenty years before the learned world knew anything about him. At twenty-five Spallanzani made translations of the ancient poets, and criticized the standard and much admired Italian translation of Homer. He brilliantly studied mathematics with his cousin, Laura Bassi, the famous woman professor of Reggio. He now skipped stones over the water in earnest, and wrote a scientific paper on the mechanics of skipping stones. He became a priest of the Catholic Church, and helped support himself by saying masses.

Despising secretly all authority, he got himself snugly into the good graces of powerful authorities, so that he might work undisturbed. Ordained a priest, supposed to be a blind fol-lower of the faith, he fell savagely to questioning everything, to taking nothing for granted—excepting the existence of God, of some sort of supreme being. At least if he questioned this he kept it—rogue that he was—strictly to himself. Before he was thirty years old he had been made professor at the University of Reggio, talking before enthusiastic classes that listened to him with saucer-eyes. Here he started his first work on the little animals, those weird new little beings that Leeu-wenhoek had discovered. He began his experiments on them as they were threatening to return to that misty unknown from which the Dutchman had dredged them up.

The little animals had got themselves involved in a strange question, in a furious fight, and had it not been for that, they might have remained curiosities for centuries, or even have been completely forgotten. This argument, over which dear friends grew to hate each other and about which professors tried to crack the skulls of priests, was briefly this: Can living things arise spontaneously, or does every living thing have to have parents? Did God create every plant and animal in the first six days, and then settle down to be Managing Director of the universe, or does He even now amuse Himself by allowing new animals to spring up in humorous ways?

In Spallanzani's time the popular side was the party that asserted that life could arise spontaneously. The great majority of sensible people believed that many animals did not have to have parents—that they might be the unhappy illegitimate children of a disgusting variety of dirty messes. Here, for example, was a supposedly sure recipe for getting yourself a good swarm of bees. Take a young bullock, kill him with a knock on the head, bury him under the ground in a standing position with his horns sticking out. Leave him there for a month, then saw off his horns—and out will fly your swarm of bees.

2

Even the scientists were on this side of the question. The English naturalist Ross announced learnedly that: "To question that beetles and wasps were generated in cow dung is to question reason, sense, and experience." Even such complicated animals as mice didn't have to have mothers or fathers—if anybody doubted this, let him go to Egypt, and there he would find the fields literally swarming with mice, begot of the mud of the River Nile—to the great calamity of the inhabitants!

Spallanzani heard all of these stories which so many important people were sure were facts, he read many more of them that were still more strange, he watched students get into brawls in excited attempts to prove that mice and bees didn't have to have fathers or mothers. He heard all of these things —and didn't believe them. He was prejudiced. Great advances in science so often start from prejudice, on ideas got not from science but straight out of a scientist's head, on notions that are only the opposite of the prevailing superstitious nonsense of the day. Spallanzani had violent notions about whether life could rise spontaneously; for him it was on the face of things absurd to think that animals—even the wee beasts of Leeuwenhoek—could arise in a haphazard way from any old thing or out of any dirty mess. There must be law and order to their birth, there must be a rime and reason! But how to prove it?

Then one night, in his solitude, he came across a little book, a simple and innocent little book, and this book told him of an entirely new way to tackle the question of how life arises. The fellow who wrote the book didn't argue with words—he just made experiments—and God! thought Spallanzani, how clear are the facts he demonstrates. He stopped being sleepy and forgot the dawn was coming, and read on. . . .

The book told him of the superstition about the generation of maggots and flies, it told of how even the most intelligent men believed that maggots and flies could arise out of putrid meat. Then—and Spallanzani's eyes nearly popped out with wonder, with excitement, as he read of a little experiment that blew up this nonsense, once and for always.

"A great man, this fellow Redi, who wrote this book," thought Spallanzani, as he took off his coat and bent his thick neck toward the light of the candle. "See how easy he settles it! He takes two jars and puts some meat in each one. He leaves one jar open and then puts a light veil over the other one. He watches—and sees flies go down into the meat in the open pot—and in a little while there are maggots there, and then new flies. He looks at the jar that has the veil over it— and there are no maggots or flies in that one at all. How easy! It is just a matter of the veil keeping the mother flies from getting at the meat. . . . But how clever, because for a thousand years people have been getting out of breath arguing about the question—and not one of them thought of doing this simple experiment that settles it in a moment."

Next morning it was one jump from the inspiring book to tackling this same question, not with flies, but with the microscopic animals. For all the professors were saying just then that though maybe flies had to come from eggs, little sub-visible animals certainly could rise by themselves.

Spallanzani began fumblingly to learn how to grow wee beasts, and how to use a microscope. He cut his hands and broke large expensive flasks. He forgot to clean his lenses and sometimes saw his little animals dimly through his fogged glasses—just as you can faintly make out minnows in the water riled up by your net. He raved at his blunders; he was not the dogged worker that Leeuwenhoek had been—but despite his impetuousness he was persistent—he must prove that these yarns about the animalcules were yarns, nothing more. But wait! "If I set out to prove something I am no real scientist— I have to learn to follow where the *facts* lead me—I have to learn to whip my prejudices. . . ." And he kept on learning to study little animals, and to observe with a patient, if not an unprejudiced eye, and gradually he taught the vanity of his ideas to bow to the hard clearness of his facts.

At this time another priest, named Needham, a devout Catholic who liked to think he could do experiments, was becoming notorious in England and Ireland, claiming that little microscopic animals were generated marvelously in mutton gravy. Needham sent his experiments to the Royal Society, and the learned Fellows deigned to be impressed.

He told them how he had taken a quantity of mutton gravy hot from the fire, and put the gravy in a bottle, and plugged the bottle up tight with a cork, so that no little animals or their eggs could possibly get into the gravy from the air. Next he even went so far as to heat the bottle and its mutton gravy in hot ashes. "Surely," said the good Needham, "this will kill any little animals or their eggs, that might remain in the flask." He put this gravy flask away for a few days, then pulled the cork—and marvel of marvels—when he examined the stuff inside with his lens, he found it swarming with animalcules.

"A momentous discovery, this," cried Needham to the Royal Society, "these little animals can only have come from the juice of the gravy. Here is a real experiment showing that life *can* come spontaneously from dead stuff!" He told them mutton gravy wasn't necessary—a soup made from seeds or almonds would do the same trick.

The Royal Society and the whole educated world were excited by Needham's discovery. Here was no Old Wives' tale. Here was hard experimental fact; and the heads of the Society got together and thought about making Needham a Fellow of their remote aristocracy of learning. But away in Italy, Spallanzani was reading the news of Needham's startling creation of little animals from mutton gravy. While he read he knit his brows, and narrowed his dark eyes. At last he snorted: "Animalcules do not arise by themselves from mutton gravy, or almond seeds, or anything else! This fine experiment is a fraud—maybe Needham doesn't know it is—but there's a nigger in the wood pile somewhere. I'm going to find it. . . ."

The devil of prejudice was talking again. Now Spallanzani began to sharpen his razors for his fellow priest—the Italian was a nasty fellow who liked to slaughter ideas of any kind that were contrary to his—he began to whet his knives, I say, for Needham. Then one night, alone in his laboratory, away from the brilliant clamor of his lectures and remote from the gay salons where ladies adored his knowledge, he felt sure he had found the loophole in Needham's experiment. He chewed his quill, he ran his hands through his shaggy hair, "Why have those little animals appeared in that hot gravy, and in those soups made from seeds?" Undoubtedly because Needham didn't heat the bottles long enough, and surely because he didn't plug them tight enough!

Here the searcher in him came forward—he didn't go to his desk to write Needham about it—instead he went to his dusty glass-strewn laboratory, and grabbed some flasks and seeds, and dusted off his microscope. He started out to test, even to defeat, if necessary, his own explanations. Needham didn't heat his soups long enough—maybe there are little animals, or their eggs, which can stand a tremendous heat, who knows? So Spallanzani took some large glass flasks, round bellied with tapering necks. He scrubbed and washed and dried them till they stood in gleaming rows on his table. Then he put seeds of various kinds into some, and peas and almonds into others, and following that poured pure water into all of them. "Now I won't only heat these soups for a short time," he cried, "but I'll boil them for an hour!" He got his fires ready—then he grunted: "But how shall I close up my flasks? Corks might not be tight enough, they might let these infinitely wee things through." He pondered. "I've got it, I'll melt the necks of my bottles shut in a flame. I'll close them with glass—nothing, no matter how small, can sneak through glass!"

So he took his shining flasks one by one, and rolled their necks gently in a hot flame till each one was fused completely shut. He dropped some of them when they got too hot—he sizzled the skin of his fingers, he swore, and got new flasks to take the smashed ones' places. Then when his flasks were all sealed and ready, "Now for some real heat," he muttered, and for tedious hours he tended his bottles, as they bumped and danced in caldrons of boiling water. One set he boiled for a few minutes only. Another he kept in boiling water for a full hour.

At last, his eyes near stuck shut with tiredness, he lifted the flasks of stew steaming from their kettles, and put them carefully away—to wait for nervous anxious days to see whether any little animals would grow in them. And he did another thing, a simple one which I almost forgot to tell you about, he made another duplicate set of stews in flasks plugged up with corks, not sealed, and after boiling these for an hour put them away beside the others.

Then he went off for days to do the thousand things that were not enough to use up his buzzing energy. He wrote letters to the famous naturalist Bonnet, in Switzerland, telling him his experiments; he played football; he went hunting and fishing. He lectured about science, and told his students not of dry technicalities only, but of a hundred things—from the marvelous wee beasts that Leeuwenhoek had found in his mouth to the strange eunuchs and the veiled multitudinous wives of Turkish harems. At last he vanished and students and professors—and ladies—asked: "Where is the Abbé Spallanzani?"

He had gone back to his rows of flasks of seed soup.

3

He went to the row of sealed flasks first, and one by one he cracked open their necks, and fished down with a slender hollow tube to get some of the soup inside them, in order to see whether any little animals at all had grown in these bottles that he had heated so long, and closed so perfectly against the microscopic creatures that might be floating in the dust of the outside air. He was not the lively sparkling Spallanzani now. He was slow, he was calm. Like some automaton, some slightly animated wooden man he put one drop of seed soup after another before his lens.

He first looked at drop after drop of the soup from the sealed flasks which had been boiled for an hour, and his long looking was rewarded by—nothing. Eagerly he turned to the bottles that had been boiled for only a few minutes, and cracked their seals as before, and put drops of the soup inside them before his lens.

"What's this?" he cried. Here and there in the gray field of his lens he made out an animalcule playing and sporting about—these weren't large microbes, like some he had seen —but they were living little animals just the same.

"Why, they look like little fishes, tiny as ants," he muttered—and then something dawned on him— "These flasks were sealed—nothing could get into them from the outside, yet here are little beings that have stood a heat of boiling water for several minutes!"

He went with nervous hands to the long row of flasks he had only stoppered with corks—as his enemy Needham had done—and he pulled out the corks, one by one, and fished in the bottles once more with his tubes. He growled excitedly, he got up from his chair, he seized a battered notebook and feverishly wrote down obscure remarks in a kind of scrawled shorthand. But these words meant that every one of the flasks which had been only corked, not sealed, was alive with little animals! Even the corked flasks which had been boiled for an hour, "were like lakes in which swim fishes of all sizes, from whales to minnows."

"That means the little animals get into Needham's flasks from the air!" he shouted. "And besides I have discovered a great new fact: living things exist that can stand boiling water and still live—you have to heat them to boiling almost an hour to kill them!"

It was a great day for Spallanzani, and though he did not know it, a great day for the world. Spallanzani had proved that Needham's theory of little animals arising spontaneously was wrong—just as the old master Redi had proved the idea was wrong that flies can be bred in putrid meat. But he had done more than that, for he had rescued the baby science of microbe hunting from a fantastic myth, a Mother Goose yarn that would have made all scientists of other kinds hold their noses at the very mention of microbe hunting as a sound branch of knowledge.

Excited, Spallanzani called his brother Nicolo, and his sister, and told them his pretty experiment. And then, bright-eyed, he told his students that life only comes from life; every living thing has to have a parent—even these wretched little animals! Seal your soup flasks in a flame, and nothing can get into them from outside. Heat them long enough, and everything, even those tough beasts that can stand boiling, will be killed. Do that, and you'll never find any living animals arising in any kind of soup—you could keep it till doomsday. Then he threw his work at Needham's head in a brilliant sarcastic paper, and the world of science was thrown into an uproar. Could Needham really be wrong? asked thoughtful men, gathered in groups under the high lamps and candles of the scientific societies of London and Copenhagen, of Paris and Berlin.

The argument between Spallanzani and Needham didn't stay in the academies among the highbrows. It leaked out through heavy doors onto the streets and crept into stylish drawing-rooms. The world would have liked to believe

Needham, for the people of the eighteenth century were cynical and gay; everywhere men were laughing at religion and denying any supreme power in nature, and they delighted in the notion that life could arise haphazardly. But Spallanzani's experiments were so clear and so hard to answer, even with the cleverest words. . . .

Meanwhile the good Needham had not been resting on his oars exactly; he was an expert at publicity, and to help his cause along he went to Paris and lectured about his mutton gravy, and in Paris he fell in with the famous Count Buffon. This count was rich; he was handsome; he loved to write about science; he believed he could make up hard facts in his head; he was rather too well dressed to do experiments. Besides he really knew some mathematics, and had translated Newton into French. When you consider that he could juggle most complicated figures, that he was a rich nobleman as well, you will agree that he certainly ought to know—without experimenting—whether little animals could come to life without fathers or mothers! So argued the godless wits of Paris.

Needham and Buffon got on famously. Buffon wore purple clothes and lace cuffs that he didn't like to muss up on dirty laboratory tables, with their dust and cluttered glassware and pools of soup spilled from accidentally broken flasks. So he did the thinking and writing, while Needham messed with the experiments. These two men then set about to invent a great theory of how life arises, a fine philosophy that every one could understand, that would suit devout Christians as well as witty atheists. The theory ignored Spallanzani's cold facts, but what would you have? It came from the brain of the great Buffon, and that was enough to upset any fact, no matter how hard, no matter how exactly recorded.

"What is it that causes these little animals to arise in mutton gravy, even after it has been heated, my Lord?" you can hear Needham asking of the noble count. Count Buffon's brain whirled in a magnificent storm of the imagination, then he answered: "You have made a great, a most momentous discovery, Father Needham. You have put your finger on the very source of life. In your mutton gravy you have uncovered the very force—it must be a force, everything is force—which creates life!"

"Let us then call it the Vegetative Force, my Lord," replied Father Needham.

"An apt name," said Buffon, and he retired to his perfumed study and put on his best suit and wrote—not from dry laboratory notes or the exact records of lenses or flasks but from his brain—he wrote, I say, about the marvels of this Vegetative Force that could make little animals out of mutton gravy and heated seed soups. In a little while Vegetative Force was on everybody's tongue. It accounted for everything. The wits made it take the place of God, and the churchmen said it was God's most powerful weapon. It was popular like a street song or an off color story—or like present day talk about relativity.

Worst of all, the Royal Society tumbled over itself to get ahead of the men in the street, and elected Needham a Fellow, and the Academy of Sciences of Paris made him an Associate. Meanwhile in Italy Spallanzani began to walk up and down his laboratory and sputter and rage. Here was a danger to science, here was ignoring of cold facts, without which science is nothing. Spallanzani was a priest of God, and God was perhaps reasonably sacred to him, he didn't argue with any one about that—but here was a pair of fellows who ignored his pretty experiments, his clear beautiful facts!

But what could Spallanzani do? Needham and Buffon had deluged the scientific world with words—they had not answered his facts, they had not shown where Spallanzani's experiment of the sealed flasks was wrong. The Italian was a fighter, but he liked to fight with facts and experiments, and here he was laying about him in this fog of big words, and hitting nothing. Spallanzani stormed and laughed and was sarcastic and bitter about this marvelous hoax, this mysterious Vegetative Force. It was the Force, prattled Needham, that had made Eve grow out of Adam's rib. It was the Force, once more, that gave rise to the remarkable worm-tree of China, which is a worm in winter, and then marvelous to say is turned by the

Vegetative Force into a tree in summer! And much more of such preposterous stuff, until Spallanzani saw the whole science of living things in danger of being upset, by this alleged Vegetative Force with which, next thing people knew, Needham would be turning cows into men and fleas into elephants.

Then suddenly Spallanzani had his chance, for Needham made an objection to one of his experiments. "Your experiment does not hold water," he wrote to the Italian, "because you have heated your flasks for an hour, and that fierce heat weakens and so damages the Vegetative Force that it can no longer make little animals."

This was just what the energetic Spallanzani was waiting for, and he forgot religion and large classes of eager students and the pretty ladies that loved to be shown through his museum. He rolled up his wide sleeves and plunged into work, not at a writing desk but before his laboratory bench, not with a pen, but with his flasks and seeds and microscopes.

4

"So Needham says heat damages the Force in the seeds, does he? Has he tried it? How can he see or feel or weigh or measure this Vegetative Force? He says it is in the seeds, well, we'll heat the seeds and see!"

Spallanzani got out his flasks once more and cleaned them. He brewed mixtures of different kinds of seeds, of peas and beans and vetches with pure water, until his work room almost ran over with flasks—they perched on high shelves, they sat on tables and chairs, they cluttered the floor so it was hard to walk around.

"Now, we'll boil a whole series of these flasks different lengths of time, and see which one generates the most little animals," he said, and then doused one set of his soups in boiling water for a few minutes, another for a half hour, another for an hour, and still another for two hours. Instead of sealing them in the flame he plugged them all up with corks —Needham said that was enough—and then he put them carefully away to see what would happen. He waited. He went off fishing and forgot to pull up his rod when a fish bit, he collected minerals for his museum, and forgot to take them home with him. He plotted for higher pay, he said masses, and studied the copulation of frogs and toads—and then disappeared once more to his dim work room with its regiments of bottles and weird machines. He waited.

If Needham were right, the flasks boiled for minutes should be alive with little animals, but the ones boiled for an hour or two hours should be deserted. He pulled out the corks one by one, and looked at the drops of soup through his lens and at last laughed with delight—the bottles that had been boiled for two hours actually had more little animals sporting about in them than the ones he had heated for a few minutes.

"Vegetative Force, what nonsense! so long as you only plug up your flasks with corks the little animals will get in from the air. You can heat your soups till you're black in the face—the microbes will get in just the same and grow, after the broth has cooled."

Spallanzani was triumphant, but then he did the curious thing that only born scientists ever do—he tried to beat his own idea, his darling theory—by experiments he honestly and shrewdly planned to defeat himself. That is science! That is the strange self-forgetting spirit of a few rare men, those curious men to whom truth is more dear than their own cherished whims and wishes. Spallanzani walked up and down his narrow work room, hands behind him, meditating—"Wait, maybe after all Needham has guessed right, maybe there is some mysterious force in these seeds that strong heat might destroy."

Then he cleaned his flasks again, and took some seeds, but instead of merely boiling them in water, he put them in a coffee-roaster and baked them till they were soot-colored cinders. Next he poured pure distilled water over them, growling: "Now if there was a Vegetative Force in those seeds, I have surely roasted it to death."

Days later when he came back to his flasks, with their soups brewed from the burned seeds, he smiled a sarcastic smile—a smile that meant squirmings for Buffon and Needham—for as one bottle after another yielded its drops of soup to his lens, every drop from every bottle was alive with wee animals that swam up and down in the liquid and went to and fro, living their funny limited little lives as gayly as any animals in the best soup made from unburned seeds. He had tried to defeat his own theory, and so trying had licked the pious Needham and the precious Buffon. They had said that heat would kill their Force so that no little animals could arise—and here were seeds charred to carbon, furnishing excellent food for the small creatures—this so-called Force was a myth! Spallanzani proclaimed this to all of Europe, which now began to listen to him.

Then he relaxed from his hard pryings into the loves and battles and deaths of little animals by making deep studies of the digestion of food in the human stomach—and to do this he experimented cruelly on himself. This was not enough, so he had to launch into weird investigations in the hot dark attic of his house, on the strange problem of how bats can keep from bumping into things although they cannot see. In the midst of this he found time to help educate his little nephews and to take care of his brother and sister, obscure beings who did not share his genius—but they were of his blood, and he loved them.

But he soon came back to the mysterious question of how life arises, that question which his religion taught him to ignore, to accept with blind faith as a miracle of the Creator. He didn't work with little animals only; instead he turned his curiosity onto larger ones, and began vast researches on the mating of toads. "What is the cause of the violent and persistent way in which the male toad holds the female?" he asked himself, and his wonder at this strange event set his ingenious brain to devising experiments of an unheard-of barbarity.

He didn't do them out of any fiendish whim to hurt the father toad—but this man must know every fact that could possibly be known about how new toads arose. What will make the toad let go this grip? And that mad priest cut off the male toad's hind legs in the midst of its copulation—but the dying animal did not relax that blind grasp to which nature drove it. Spallanzani mused over his bizarre experiment. "This persistence of the toad," he said, "is due less to his obtuseness of feeling than to the vehemence of his passion."

In his sniffing search for knowledge which let him stop at nothing, he was led by an instinct that drove him into heartless experiments on animals—but it made him do equally cruel and fantastic tests on himself. He studied the digestion of food in the stomach, he gulped down hollowed-out blocks of wood with meat inside them, then tickled his throat and made himself vomit them up again so that he could find out what had happened to the meat inside the blocks. He kept insanely at this self-torture, until, as he admitted at last, a horrid nausea made him stop the experiments.

Spallanzani held immense correspondences with half the doubters and searchers of Europe. By mail he was a great friend of that imp, Voltaire. He complained that there were few men of talent in Italy, the air was too humid and foggy— he became a leader of that impudent band of scientists and philosophers who unknowingly prepared the bloodiest of revolutions while they tried so honestly to find truth and establish happiness and justice in the world. These men believed that Spallanzani had spiked once for all that nonsense about animals—even the tiniest ones—arising spontaneously. Led by Voltaire they cracked vast jokes about the Vegetative Force and its parents, the pompous Buffon and his laboratory boy, Father Needham.

"But there is a Vegetative Force," cried Needham, "a mysterious something—I'll admit you can't see it or weigh it— that can make life arise out of gravy or soup or out of nothing at all, perhaps. Maybe it can stand all of that roasting that Spallanzani applies to it, but what it needs particularly is a very elastic air to help it. And when Spallanzani boils his flasks for an hour, he hurts the elasticity of the air inside the flasks!"

Spallanzani was up in arms in a moment, and bawled for

Needham's experiments. "Has he heated air to see if it got less elastic?" The Italian waited for experiments—and got only words. "Then I'll have to test it out myself," he said, and once again he put seeds in rows of flasks and sealed off their necks in a flame—and boiled them for an hour. Then one morning he went to his laboratory, and cracked off the neck of one of his bottles. . . .

He cocked his ear—he heard a little wh-i-s-s-s-s-t. "What's this," he muttered, and grabbed another bottle and cracked off its neck, holding his ear close by. Wh-i-s-s-st! There it was again. "That means the air is coming out of my bottle, or going into it," he cried, and he lighted a candle and ingeniously held it near the neck of a third flask as he cracked the seal.

The flame sucked inward toward the opening.

"The air's going in—that means the air in the bottle is less elastic than the air outside, that means maybe Needham is right!"

For a moment Spallanzani had a queer feeling at the pit of his stomach, his forehead was wet with nervous sweat, his world tottered around him. . . . Could that fool Needham have made a lucky stab, a clever guess about what heat did to air in sealed up flasks? Could this windbag knock out all of this careful finding of facts, which had taken so many years of hard work? For days Spallanzani went about troubled, and snapped at students to whom before he had been gentle, and tried to comfort himself by reciting Dante and Homer—and this only made him more grumpy. A relentless torturing imp pricked at him and this imp said: "Find out why the air rushes into your flasks when you break the seals—it may not have anything to do with elasticity." The imp woke him up in the night, it made him get tangled up in his masses. . . .

Then like a flash of lightning the explanation came to him and he hurried to his work bench—it was covered with broken flasks and abandoned bottles and its muddled disarray told his discouragement—he reached into a cupboard and took out one of his flasks. He was on the track, he would show Needham was wrong, and even before he had proved it he stretched himself with a heave of relief—so sure was he that the reason for the little whistling of air had come to him. He looked at the flasks, then smiled and said, "All the flasks that I have been using have fairly wide necks. When I seal them in the flame it takes a lot of heat to melt the glass till the neck is shut off —all that heat drives most of the air out of the bottle before it's sealed up. No wonder the air rushes in when I crack the seal!"

He saw that Needham's idea that boiling water outside the flask damaged the elasticity of the air inside was nonsense, nothing less. But how to prove this, how to seal up the flasks without driving out the air? His devilish ingenuity came to help him, and he took another flask, put seeds into it, and filled it partly with pure water. Then he rolled the neck of the bottle around in a hot flame until it melted down to a tiny narrow opening—very, very narrow, but still open to the air outside. Next he let the flask cool—now the air inside must be the same as the air outside—then he applied a tiny flame to the now almost needle-fine opening. In a jiffy the flask was sealed—without expelling any of the air from the inside. Content, he put the bottle in boiling water and watched it bump and dance in the kettle for an hour and while he watched he recited verses and hummed gay tunes. He put the flask away for days, then one morning, sure of his result, he came to his laboratory to open it. He lighted a candle; he held it close to the flask neck; carefully he broke the seal—wh-i-s-s-s-t! But the flame blew *away* from the flask this time—the elasticity of the air inside the flask was greater than that outside!

All of the long boiling had not damaged the air at all—it was even more elastic than before—and elasticity was what Needham said was necessary for his wonderful Vegetative Force. The air in the flask was super-elastic, but fishing drop after drop of the soup inside, Spallanzani couldn't find a single little animal. Again and again, with the obstinacy of a Leeuwenhoek, he repeated the same experiment. He broke flasks and spilled boiling water down his shirt-front, he seared his hands, he made vast tests that had to be done over—but always he confirmed his first result.

5

Triumphant he shouted his last experiment to Europe, and Needham and Buffon heard it, and had to sit sullenly amid the ruins of their silly theory, there was nothing to say—Spallanzani had spiked their guns with a simple fact. Then the Italian sat down to do a little writing himself. A virtuoso in the laboratory, he was a fiend with his quill, when once he was sure his facts had destroyed Needham's pleasant myth about life arising spontaneously. Spallanzani was sure now that even the littlest beasts had to come—always—from beasts that had lived before. He was certain too, that a wee microbe always remained a microbe of the same kind that its parents had been, just as a zebra doesn't turn into a giraffe, or have musk-oxen for children, but always stays a zebra—and has zebra babies.

"In short," shouted Spallanzani, "Needham is wrong, and I have proved that there is a law and order in the science of animals, just as there is in the working of the stars."

Then he told the muddle that Needham would have turned the science of little animals into—if good facts hadn't been found to beat him. What animals this weird Vegetative Force could make—what tricks it could do—if it had only existed! "It could make," said Spallanzani, "a microscopic animal found sometimes in infusions, which like a new Protean, ceaselessly changes its form, appearing now as a body thin as a thread, now in an oval or spherical form, sometimes coiled like a serpent, adorned with rays and armed with horns. This remarkable animal furnishes Needham an example, to explain easily how the Vegetative Force produces now a frog and again a dog, sometimes a midge and at others an elephant, to-day a spider and to-morrow a whale, this minute a cow and the next a man."

So ended Needham—and his Vegetative Force. It became

comfortable to live once more; you felt sure there was no mysterious sinister Force sneaking around waiting to change you into a hippopotamus.

Spallanzani's name glittered in all the universities of Europe; the societies considered him the first scientist of the day; Frederick the Great wrote long letters to him and with his own hand made him a member of the Berlin Academy; and Frederick's bitter enemy, Maria Theresa, Empress of Austria, put it over the great king by offering Spallanzani the job of professor in her ancient and run-down University of Pavia, in Lombardy. A pompous commission came, a commission of eminent Privy Councillors weighed down with letters and Imperial Seals and begged Spallanzani to put this defunct college on its feet. There were vast interminable arguments and bargainings about salary—Spallanzani always knew how to feather his nest—bargains that ended in his taking the job of Professor of Natural History and Curator of the Natural History Cabinet of Pavia.

Spallanzani went to the Museum, the Natural History Cabinet, and found that cupboard bare. He rolled up his sleeves, he lectured about everything, he made huge public experiments and he awed his students because his deft hands always made these experiments turn out successfully. He sent here and there for an astounding array of queer beasts and strange plants and unknown birds—to fill up the empty Cabinet. He climbed dangerous mountains himself and brought back minerals and precious ores; he caught hammer-head sharks and snared gay-plumed fowl; he went on incredible collecting expeditions for his museum—and to work off that tormenting energy that made him so fantastically different from the popular picture of a calm scientist. He was a Roosevelt with all of Teddy's courage and appeal to the crowd, but with none of Teddy's gorgeous inaccuracy.

In the intervals of this hectic collecting and lecturing he shut himself in his laboratory with his stews and his microscopic animals, and made long experiments to show that these beasts obey nature's laws, just as men and horses and elephants are forced to follow them. He put drops of stews swarming with microbes on little pieces of glass and blew tobacco smoke at them and watched them eagerly with his lens. He cried out his delight as he saw them rush about trying to avoid the irritating smoke. He shot electric sparks at them and wondered at the way the little animals "became giddy" and spun about, and quickly died.

"The seeds or eggs of the little animals may be different from chicken eggs or frog's eggs or fish eggs—they may stand the heat of boiling water in my sealed flasks—but otherwise these little creatures are really no different from other animals!" he cried. Then just after that he had to take back his confident words. . . .

"Every beast on earth needs air to live, and I am going to show just how *animal* these little animals are by putting them in a vacuum—and watching them die," said Spallanzani to himself, alone one day in his laboratory. He cleverly drew out some very thin tubes of glass, like the ones Leeuwenhoek had used to study his little animals. He dipped the tube into a soup that swarmed with his microbes; the fluid rushed up into the hair-fine pipe. Then Spallanzani sealed off one end of it, and ingeniously tied the other end to a powerful vacuum pump, and set the pump going, and stuck his lens against the thin wall of the tube. He expected to see the wee animals stop waving the "little arms which they were furnished to swim with"; he expected them to get giddy and then stop moving. . . .

The pump chugged on—and nothing whatever happened to the microbes. They went nonchalantly about their business and did not seem to realize there was such a thing as life-maintaining air! They lived for days, for weeks—and Spallanzani did the experiment again and again, trying to find something wrong with it. This was impossible—nothing can live without air—how the devil do these beasts breathe? He wrote his amazement in a letter to his friend Bonnet:

"The nature of some of these animalcules is astonishing! They are able to exercise in a vacuum the functions they use in free air. They make all of their courses, they go up and down in the liquid, they even multiply for several days in this vacuum. How wonderful this is! For we have always believed there is no living being that can live without the advantages air offers it."

Spallanzani was very proud of his imagination and his quick brain and he was helped along in this conceit by the flattery and admiration of students and intelligent ladies and learned professors and conquering kings. But he was an experimenter too—he was really an experimenter first, and he bent his head humbly when a new fact defeated one of the brilliant guesses of his brain.

Meanwhile this man who was so rigidly honest in his experiments, who would never report anything but the truth of what he found amid the smells and poisonous vapors and shining machines of his laboratory, this superbly honest scientist, I say, was planning low tricks to increase his pay as Professor at Pavia. Spallanzani, the football player, the climber of mountains and explorer, this Spallanzani whined to the authorities at Vienna about his feeble health—the fogs and vapors of Pavia were like to make him die, he said. To keep him the Emperor had to increase his pay and double his vacations. Spallanzani laughed and cynically called his lie a political gesture! He always got everything he wanted. He got truth by dazzling experiments and close observation and insane patience; he obtained money and advancement by work—and by cunning plots and falsehoods; he received protection from religious persecution by becoming a priest!

Now, as he grew older, he began to hanker for wild researches in regions remote from his little laboratory. He must visit the site of ancient Troy whose story thrilled him so; he must see the harems and slaves and eunuchs, which to him were as much a part of natural history as his bats and toads and little animals of the seed infusions. He pulled wires, and at last the Emperor Joseph gave him a year's leave of absence and the money for a trip to Constantinople—for his failing health, which had never been more superb.

So Spallanzani put his rows of flasks away and locked his laboratory and said a dramatic and tearful good-by to his students; on the journey down the Mediterranean he got frightfully sea-sick, he was shipwrecked—but didn't forget to try to save the specimens he had collected on some islands. The Sultan wined and dined him, the doctors of the seraglios let him study the customs of the beauteous concubines . . . and afterward, good eighteenth century European that he was, Spallanzani told the Turks that he admired their hospitality and their architecture, but detested their custom of slavery and their hopeless fatalistic view of life. . . .

"We Westerners, through this new science of ours, are going to conquer the seemingly unavoidable, the apparently eternal torture and suffering of man," you can imagine him telling his polite but stick-in-the-mud Oriental friends. He believed in an all powerful God, but while he believed, the spirit of the searcher, the fact finder, flashed out of his eye, burdened all his thought and talk, forced him to make excuses for God by calling him Nature and the Unknown, compelled him to show that he had appointed himself first assistant to God in the discovery and even the conquering of this unknown Nature.

After many months he returned overland through the Balkan Peninsula, escorted by companies of crack soldiers, entertained by Bulgarian dukes and Wallachian Hospodars. At last he came to Vienna, to pay his respects to his boss and patron, the Emperor Joseph II—it was the dizziest moment, so far as honors went, of his entire career. Drunk with success, he thought, you may imagine, of how all of his dreams had come true, and then—

6

While Spallanzani was on his triumphant voyage a dark cloud gathered away to the south, at his university, the school at Pavia that he had done so much to bring back to life. For years the other professors had watched him take their students away from them, they had watched—and ground their tusks and sharpened their razors—and waited.

Spallanzani by tireless expeditions and through many fatigues and dangers had made the once empty Natural History Cabinet the talk of Europe. Besides he had a little private collection of his own at his old home in Scandiano. One day, Canon Volta, one of his jealous enemies, went to Scandiano and by a trick got into Spallanzani's private museum; he sniffed around, then smiled an evil grin—here were some jars, and there a bird and in another place a fish, and all of them were labeled with the red tags of the University museum of Pavia! Volta sneaked away hidden in the dark folds of his cloak, and on the way home worked out his malignant plans to cook the brilliant Spallanzani's goose; and just before Spallanzani got home from Vienna, Volta and Scarpa and Scopoli let hell loose by publishing a tract and sending it to every great man and society in Europe, and this tract accused Spallanzani of the nasty crime of stealing specimens from the University of Pavia and hiding them in his own little museum at Scandiano.

His bright world came down around his ears; in a moment he saw his gorgeous career in ruins; in hideous dreams he heard the delighted cackles of men who praised and envied him; he pictured the triumph of men whom he had soundly licked with his clear facts and experiments—he imagined even the return to life of that fool Vegetative Force. . . .

But in a few days he came back on his feet, the center of a dreadful scandal, it is true, but on his feet with his back to the wall ready to face his accusers. Gone now was the patient hunter of microbes and gone the urbane correspondent of Voltaire. He turned into a crafty politician, he demanded an investigating committee and got it, he founded Ananias Clubs, he fought fire with fire.

He returned to Pavia and on his way there I wonder what his thoughts were—did he see himself slinking into the town, avoided by old admirers and a victim of malignant hissing whispers? Possibly, but as he got near the gates of Pavia a strange thing happened—for a mob of adoring students came out to meet him, told him they would stick by him, escorted him with yells of joy to his old lecture chair. The once self-sufficient, proud man's voice became husky—he blew his nose—he could only stutteringly tell them what their devotion meant to him.

Then the investigating committee had him and his accusers appear before it, and knowing Spallanzani as you already do, you may imagine the shambles that followed! He proved to the judges that the alleged stolen birds were miserably stuffed, draggle-feathered creatures which would have disgraced the cabinet of a country school—they had been merely pitched out. He had traded the lost snakes and the armadillo to other museums and Pavia had profited by the trade; not only so, but Volta, his chief accuser, had himself stolen precious stones from the museum and given them to his friends. . . .

The judges cleared him of all guilt—though it is to-day not perfectly sure that he wasn't a little guilty; Volta and his co-plotters were fired from the University, and all parties, including Spallanzani, were ordered by the Emperor to stop their deplorable brawling and shut up—this thing was getting to be a smell all over Europe—students were breaking up the classroom furniture about it, and other universities were snickering at such an unparalleled scandal. Spallanzani took a last crack at his routed enemies; he called Volta a perfect bladder full of wind and invented hideous and unprintably improper names for Scarpa and Scopoli; then he returned peacefully to his microbe hunting.

Many times in his long years of looking at the animalcules he had wondered how they multiplied. Often he had seen two of the wee beasts stuck together, and he wrote to Bonnet: "When you see two individuals of any animal kind united, you naturally think they are engaged in reproducing themselves." But were they? He jotted his observations down in old note-books and made crude pictures of them, but, impetuous as he was in many things, when it came to experiments or drawing conclusions—he was almost as cagy as old Leeuwenhoek had been.

Bonnet told Spallanzani's perplexity about the way little animals multiplied to his friend, the clever but now unknown de Saussure. And this fellow turned his sharp eye through his clear lenses onto the breeding habits of animalcules. In a short while he wrote a classic paper, telling the fact that when you see two of the small beasts stuck together, they haven't come together to breed. On the contrary—marvelous to say—these coupled beasts are nothing more nor less than an old animalcule which is dividing into two parts, into two new little animals! This, said de Saussure, was the only way the microbes ever multiplied—the joys of marriage were unknown to them!

Reading this paper, Spallanzani rushed to his microscope hardly believing such a strange event could be so—but careful looking showed that de Saussure was right. The Italian wrote the Swiss a fine letter congratulating him; Spallanzani was a fighter and something of a plotter; he was infernally ambitious and often jealous of the fame of other men, but he lost himself in his joy at the prettiness of de Saussure's sharp observations. Spallanzani and these naturalists of Geneva were bound by a mysterious cement—a realization that the work of finding facts and fitting facts together to build the high cathedral of science is greater than any single finder of facts or mason of facts. They were the first haters of war—the first citizens of the world, the first genuine internationalists.

Then Spallanzani was forced into one of the most devilishly ingenious researches of his life. He was forced into this by his friendship for his pals in Geneva and by his hatred of another piece of scientific claptrap almost as bad as the famous Vegetative Force. An Englishman named Ellis wrote a paper saying de Saussure's observations about the little animals splitting into two were all wrong. Ellis admitted that the little beasts might occasionally break into two. "But that," cried Ellis, "doesn't mean they are multiplying! It simply means," he said, "that one little animal, swimming swiftly along in the water, bangs into another one amidships—and breaks him in half! That's all there is to de Saussure's fine theory.

"What is more," Ellis went on, "little animals are born from each other just as larger beasts come from their mothers. When I look carefully with my microscope, I can actually see young ones inside the old ones, and looking still more closely—you may not believe it—I can see grandchildren inside these young ones."

"Rot!" thought Spallanzani. All this stuff smelled very fishy to him, but how to show it wasn't true, and how to show that animalcules multiplied by breaking in two?

He was first of all a hard scientist, and he knew that it was one thing to say Ellis was feeble-minded, but quite another to *prove* that the little animals didn't bump into each other and so knock each other apart. In a moment the one way to decide it came to him— "All I have to do," he meditated, "is to get one little beast off by itself, away from every other one where nothing whatever can bump into it—and then just sit and watch through the microscope to see if it breaks into two." That was the simple and the only way to do it, no doubt, but how to get one of these infernally tiny creatures away from his swarms of companions? You can separate one puppy from a litter, or even a little minnow from its myriads of brothers and sisters. But you can't reach in with your hands and take one animalcule by the tail—curse it—it is a million times too small for that.

Then this Spallanzani, this fellow who reveled in gaudy celebrations and vast enthusiastic lecturings, this hero of the crowd, this magnifico, crawled away from all his triumphs and pleasures to do one of the cleverest and most marvelously ingenious pieces of patient work in his hectic life. He did no less a thing than to invent a sure method of getting *one* animalcule—a few twenty-five thousandths of an inch long— a living animalcule, off by itself.

He went to his laboratory and carefully put a drop of seed soup swarming with animalcules on a clean piece of crystal glass. Then with a clean hair-fine tube he put a drop of pure distilled water—that had not a single little animal in it—on the same glass, close to the drop that swarmed with microbes.

"Now I shall trap one," he muttered, as he trained his lens on the drop that held the little animals. He took a fine clean needle, he stuck it carefully into the drop of microbe soup— and then made a little canal with it across to the empty water drop. Quickly he turned his lens onto the passageway between the two drops, and grunted satisfaction as he saw the wriggling cavorting little creatures begin to drift through this little canal. He grabbed for a little camel's-hair brush— "There! there's one of the wee ones—just one, in the water drop!" Deftly he flicked the little brush across the small canal, wiping it out, so cutting off the chance of any other wee beast getting into the water drop to join its lonely little comrade.

"God!" he cried. "I've done it—no one's ever done this before—I've got one animalcule all by himself; now nothing can bump him, now we'll see if he'll turn into two new ones!" His lens hardly quivered as he sat with tense neck and hands and arms, back bent, eye squinting through the glass at the drop with its single inhabitant. "How tiny he is," he thought —"he is like a lone fish in the spacious abysses of the sea."

Then a strange sight startled him, not less dramatic for its unbelievable littleness. The beast—it was shaped like a small rod—began to get thinner and thinner in the middle. At last the two parts of it were held together by the thickness of a spider web thread, and the two thick halves began to wriggle desperately—and suddenly they jerked apart. There they were, two perfectly formed, gently gliding little beasts, where there had been one before. They were a little shorter but otherwise they couldn't be told from their parent. Then, what was more marvelous to see, these two children of the first one in a score of minutes split up again—and now there were four where there had been one!

Spallanzani did this ingenious trick a dozen times and got the same result and saw the same thing; and then he descended on the unlucky Ellis like a ton of brick and flattened into permanent obscurity Ellis and his fine yarn about the children and the grandchildren inside the little animals. Spallanzani was sniffish, he condescended, he advised, he told Ellis to go back to school and learn his a b c's of microbe hunting. He hinted that Ellis wouldn't have made his mistake if he'd read the fine paper of de Saussure carefully, instead of inventing preposterous theories that only cluttered up the hard job of getting genuine new facts from a stingy Nature.

A scientist, a really original investigator of nature, is like a writer or a painter or a musician. He is part artist, part cool searcher. Spallanzani told himself stories, he conceived himself the hero of a new epic exploration, he compared himself—in his writings even—to Columbus and Vespucci. He told of that mysterious world of microbes as a new universe, and thought of himself as a daring explorer making first groping expeditions along its boundaries only. He said nothing about the possible deadliness of the little animals—he didn't like to engage, in print, in wild speculations—but his genius whispered to him that the fantastic creatures of this new world were of some sure but yet unknown importance to their big brothers, the human species. . . .

7

Early in the year 1799, as Napoleon started thoroughly smashing an old world to pieces, and just as Beethoven was knocking at the door of the nineteenth century with the first of his mighty symphonies, war-cries of that defiant spirit of which Spallanzani was one of the chief originators—in the year 1799, I say, the great microbe hunter was struck with apoplexy. Three days later he was poking his energetic and irrepressible head above the bedclothes, reciting Tasso and Homer to the amusement and delight of those friends who had come to watch him die. But though he refused to admit it, this, as one of his biographers says, was his *Canto di Cigno*, his swan song, for in a few days he was dead.

Great Egyptian kings kept their names alive for posterity by having the court undertaker embalm them into expensive and gorgeous mummies. The Greeks and Romans had their likenesses wrought into dignified statues. Paintings exist of a hundred other distinguished men. What is left for us to see of the marvelous Spallanzani?

In Pavia there is a modest little bust of him and in the museum near by, if you are interested, you may see—his bladder. What better epitaph could there be for Spallanzani? What relic could more perfectly suggest the whole of his passion to find truth, that passion which stopped at nothing, which despised conventions, which laughed at hardship, which ignored bad taste and the feeble pretty fitness of things?

He knew his bladder was diseased. "Well, have it out after I'm dead," you can hear him whisper as he lay dying. "Maybe you'll find an astonishing new fact about diseased bladders." That was the spirit of Spallanzani. This was the very soul of that cynical, sniffingly curious, coldly reasoning century of his—the century that discovered few practical things—but the same century that built the high clean house for Faraday and Pasteur, for Arrhenius and Emil Fischer and Ernest Rutherford to work in.

3

Pasteur

Microbes Are a Menace!

1

In 1831, thirty-two years after the magnificent Spallanzani died, microbe hunting had come to a standstill once more. The sub-visible animals were despised and forgotten while other sciences were making great leaps ahead; clumsy horribly coughing locomotives were scaring the horses of Europe and America; the telegraph was getting ready to be invented. Marvelous microscopes were being devised, but no man had come to squint through these machines—no man had come to prove to the world that miserable little animals could do useful work which no complicated steam engine could attempt; there was no hint of the somber fact that these wretched microbes could kill their millions of human beings mysteriously and silently, that they were much more efficient murderers than the guillotine or the cannon of Waterloo.

On a day in October in 1831, a nine-year-old boy ran frightened away from the edge of a crowd that blocked the door of the blacksmith shop of a village in the mountains of eastern France. Above the awed excited whispers of the people at the door this boy had heard the crackling "s-s-s-s-z" of a white hot iron on human flesh, and this terrifying sizzling had been followed by a groan of pain. The victim was the farmer Nicole. He had just been mangled by a mad wolf that charged howling, jaws dripping poison foam, through the streets of the village. The boy who ran away was Louis Pasteur, son of a tanner of Arbois and great-grandson of a serf of the Count of Udressier.

Days and weeks passed and eight victims of the mad wolf died in the choking throat-parched agonies of hydrophobia. Their screams rang in the ears of this timid—some called him stupid—boy; and the iron that had seared the farmer's wound burned a deep scar in his memory.

"What makes a wolf or a dog mad, father—why do people die when mad dogs bite them?" asked Louis. His father the tanner was an old sergeant of the armies of Napoleon. He had seen ten thousand men die from bullets, but he had no notion of why people die from disease. "Perhaps a devil got into the wolf, and if God wills you are to die, you will die, there is no help for it," you can hear the pious tanner answer. That answer was as good as any answer from the wisest scientist or the most expensive doctor in the world. In 1831 no one knew what caused people to die from mad dog bites—the cause of all disease was completely unknown and mysterious.

I am not going to try to make believe that this terrible event made the nine-year-old Louis Pasteur determine to find out the cause and cure of hydrophobia some day—that would be very romantic—but it wouldn't be true. It is true though that he was more scared by it, haunted by it for a longer time, brooded over it more, that he smelled the burned flesh and heard the screams a hundred times more vividly than an ordinary boy would—in short, he was of the stuff of which artists are made; and it was this stuff in him, as much as his science, that helped him to drag microbes out of that obscurity into which they had passed once more, after the gorgeous Spallanzani died. Indeed, for the first twenty years of his life he showed no signs at all of becoming a great searcher. This Louis Pasteur was only a plodding, careful boy whom nobody noticed particularly. He spent his playtime painting pictures of the river that ran by the tannery, and his sisters posed for him until their necks grew stiff and their backs ached grievously; he painted curiously harsh unflattering pictures of his mother—they didn't make her look pretty, but they looked like his mother. . . .

Meanwhile it seemed perfectly certain that the little animals were going to be put permanently on the shelf along with the dodo and other forgotten beasts. The Swede Linnæus, most enthusiastic pigeonholer, who toiled at putting all living things in a neat vast card catalogue, threw up his hands at the very idea of studying the wee beasts. "They are too small, too confused, no one will ever know anything exact about them, we will simply put them in the class of Chaos!" said Linnæus. They were only defended by the famous round-faced German Ehrenberg who had immense quarrels—in moments when he wasn't crossing oceans or receiving medals—futile quarrels about whether the little animals had stomachs, strange arguments about whether they were really complete little animals or only parts of larger animals; or whether perchance they might be little vegetables instead of little animals.

Pasteur kept plugging at his books though, and it was while he was still at the little college of Arbois that the first of his masterful traits began to stick out—traits good and bad, that made him one of the strangest mixtures of contradictions that ever lived. He was the youngest boy at the college, but he wanted to be a monitor; he had a fiery ambition to teach other boys, particularly to run other boys. He became a monitor. Before he was twenty he had become a kind of assistant teacher in the college of Bezançon, and here he worked like the devil and insisted that everybody else work as hard as he worked himself; he preached in long inspirational letters to his poor sisters—who, God bless them, were already trying their best—

"To *will* is a great thing, dear sisters," he wrote, "for Action and Work usually follow Will, and almost always Work is accompanied by Success. These three things, Work, Will, Success, fill human existence. Will opens the door to success both brilliant and happy; Work passes these doors, and at the end of the journey Success comes to crown one's efforts."

When he was seventy his sermons had lost their capital letters, but they were exactly the same kind of simple earnest sermons.

His father sent him up to Paris to the Normal School and there he resolved to do great things, but he was carried away by a homesickness for the smell of the tannery yard and he came back to Arbois abandoning his high ambition. . . . In another year he was back at the same school in Paris and this time he stuck at it; and then one day he passed in a tear-stained trance out of the lecture room of the chemist Dumas. "What a science is chemistry," he muttered, "and how marvelous is the popularity and glory of Dumas." He knew then that he was going to be a great chemist too; the misty gray streets of the Latin Quarter dissolved into a confused and frivolous world that chemistry alone could save. He had left off his painting but he was still the artist.

Presently he began to make his first stumbling independent researches with stinking bottles and rows of tubes filled with gorgeous colored fluids. His good friend Chappuis, a mere student of philosophy, had to listen for hours to Pasteur's lectures on the crystals of tartaric acid, and Pasteur told Chappuis: "It is sad that you are not a chemist too." He would have made all students chemists just as forty years later he tried to turn all doctors into microbe hunters.

Just then, as Pasteur was bending his snub nose and broad forehead over confused piles of crystals, the sub-visible living microbes were beginning to come back into serious notice, they were beginning to be thought of as important serious fellow creatures, just as useful as horses or elephants, by two lonely searchers, one in France and one in Germany. A modest but original Frenchman, Cagniard de la Tour, in 1837 poked round in beer vats of breweries. He dredged up a few foamy drops from such a vat and looked at them through a microscope and noticed that the tiny globules of the yeasts he found in them sprouted buds from their sides, buds like seeds sprouting. "They are alive then, these yeasts, they multiply like other creatures," he cried. His further searchings made him see that no brew of hops and barley ever changed into beer without the presence of the yeasts, living growing yeasts. "It must be their *life* that changes barley into alcohol," he meditated, and he wrote a short clear paper about it. The world refused to get excited about this fine work of the wee yeasts—Cagniard was no propagandist, he had no press agent to offset his own modesty.

In the same year in Germany Doctor Schwann published a short paper in long sentences, and these muddy phrases told a bored public the exciting news that meat only becomes putrid when sub-visible animals get into it. "Boil meat thoroughly and put it in a clean bottle and lead air into it that has passed through red-hot pipes—the meat will remain perfectly fresh for months. But in a day or two after you remove the stopper and let in ordinary air, with its little animals, the meat will begin to smell dreadfully; it will teem with wriggling, cavorting creatures a thousand times smaller than a pinhead—it is these beasts that make meat go bad."

How Leeuwenhoek would have opened his large eyes at this! Spallanzani would have dismissed his congregation and rushed from his masses to his laboratory; but Europe hardly looked up from its newspapers, and young Pasteur was getting ready to make his own first great chemical discovery.

When he was twenty-six years old he made it. After long peerings at heaps of tiny crystals he discovered that there are four distinct kinds of tartaric acid instead of two; that there are a variety of strange compounds in nature that are exactly alike—excepting that they are mirror-images of each other. When he stretched his arms and straightened up his lame back and realized what he had done, he rushed out of his dirty dark little laboratory into the hall, threw his arms around a young physics assistant—he hardly knew him—and took him out under the thick shade of the Gardens of the Luxembourg. There he poured mouthfuls of triumphant explanation at him—he must tell some one. He wanted to tell the world!

2

In a month he was praised by gray-haired chemists and became the companion of learned men three times his age. He was made professor at Strasbourg and in the off moments of researches he determined to marry the daughter of the dean. He didn't know if she cared for him but he sat down and wrote her a letter that he knew must make her love him:

"There is nothing in me to attract a young girl's fancy," he wrote, "but my recollections tell me that those who have known me very well have loved me very much."

So she married him and became one of the most famous and long-suffering and in many ways one of the happiest wives in history—and this story will have more to tell about her.

Now the head of a house, Pasteur threw himself more furiously into his work; forgetting the duties and chivalries of a bridegroom, he turned his nights into days. "I am on the verge of mysteries," he wrote, "and the veil is getting thinner and thinner. The nights seem to me too long. I am often scolded by Madame Pasteur, but I tell her I shall lead her to fame." He continued his work on crystals; he ran into blind alleys, he did strange and foolish and impossible experiments, the kind a crazy man might devise—and the kind that turn a crazy man into a genius when they come off. He tried to change the chemistry of living things by putting them between huge magnets. He devised weird clockworks that swung plants back and forward, hoping so to change the mysterious molecules that formed these plants into mirror images of themselves. . . . He tried to imitate God: he tried to change species!

Madame Pasteur waited up nights for him and marveled at him and believed in him, and she wrote to his father: "You know that the experiments he is undertaking this year will give us, if they succeed, a Newton or a Galileo!" It is not clear whether good Madame Pasteur formed this so high opinion of her young husband by herself. . . . At any rate, truth, that will o' the wisp, failed him this time—his experiments didn't come off.

Then Pasteur was made Professor and Dean of the Faculty of Sciences in Lille and there he settled down in the Street of the Flowers, and it was here that he ran, or rather stumbled for the first time, upon microbes; it was in this good solid town of distillers and sugar-beet raisers and farm implement dealers that he began his great campaign, part science, part drama and romance, part religion and politics, to put microbes on the map. It was from this not too interesting middle sized city— never noted for learning—that he splashed up a great wave of excitement about microbes that rocked the boat of science for thirty years. He showed the world how important microbes were to it, and in doing this he made enemies and worshipers; his name filled the front pages of newspapers and he received challenges to duels; the public made vast jokes about his precious microbes while his discoveries were saving the lives of countless women in childbirth. In short it was here he hopped off in his flight to immortality.

When he left Strasbourg truth was tricking him and he was confused. He came to Lille and fairly stumbled on to the road to fame—by offering help to a beet-sugar distiller.

When Pasteur settled in Lille he was told by the authorities that highbrow science was all right—

"But what we want, what this enterprising city of Lille wants most of all, professor," you can hear the Committee of business men telling him, "is a close coöperation between your science and our industries. What we want to know is—does science pay? Raise our sugar yield from our beets and give us a bigger alcohol output, and we'll see you and your laboratory are taken care of."

Pasteur listened politely and then proceeded to show them the stuff he was made of. He was much more than a man of science! Think of a committee of business men asking Isaac Newton to show them how his laws of motion were going to help their iron works! That shy thinker would have thrown up his hands and set himself to studying the meaning of the prophecies of the Book of Daniel at once. Faraday would have gone back to his first job as a bookbinder's apprentice. But Pasteur was no shrinking flower. A child of the nineteenth century, he understood that science had to earn its bread and butter, and he started to make himself popular with everybody by giving thrilling lectures to the townspeople on science:

"Where in your families will you find a young man whose curiosity and interest will not immediately be awakened when you put into his hands a potato, and when with that potato he may produce sugar, and with that sugar alcohol, and with that alcohol ether and vinegar?" he shouted enthusiastically one evening to an audience of prosperous manufacturers and their wives. Then one day Mr. Bigo, a distiller of alcohol from sugar beets, came to his laboratory in distress. "We're having trouble with our fermentations, Professor," he complained; "we're losing thousands of francs every day. I wonder if you could come over to the factory and help us out?" said the good Bigo.

Bigo's son was a student in the science course and Pasteur hastened to oblige. He went to the distillery and sniffed at the vats that were sick, that wouldn't make alcohol; he fished up some samples of the grayish slimy mess and put them in bottles to take to his laboratory—and he didn't fail to take some of the beet pulp from the healthy foamy vats where good amounts of alcohol were being made. Pasteur had no idea he could help Bigo, he knew nothing of how sugar ferments into alcohol—indeed, no chemist in the world knew anything about it. He got back to his laboratory, scratched his head, and decided to examine the stuff from the healthy vats first. He put some of this stuff—a drop of it—before his microscope, maybe with an aimless idea of looking for crystals, and he found this drop was full of tiny globules, much smaller than any crystal, and these little globes were yellowish in color, and their insides were full of a swarm of curious dancing specks.

"What can these things be," he muttered. Then suddenly he remembered—

"Of course, I should have known—these are the yeasts you find in all stews that have sugar which is fermenting into alcohol!"

He looked again and saw the wee spheres alone; he saw some in bunches, others in chains, and then to his wonder he came on some with queer buds sprouting from their sides— they looked like sprouts on infinitely tiny seeds.

"Cagniard de la Tour is right. These yeasts are alive. It must be the yeasts that change beet sugar into alcohol!" he cried. "But that doesn't help Mr. Bigo—what on earth can be the matter with the stuff in the sick vats?" He grabbed for the bottle that held the stuff from the sick vat, he sniffed at it, he peered at it with a little magnifying glass, he tasted it, he dipped little strips of blue paper in it and watched them turn red. . . . Then he put a drop from it before his microscope and looked. . . .

"But there are no yeasts in this one; where are the yeasts? There is nothing here but a mass of confused stuff—what is it, what does this mean?" He took the bottle up again and brooded over it with an eye that saw nothing—till at last a different, a strange look of the juice forced its way up into his wool-gathering thoughts. "Here are little gray specks sticking to the walls of the bottle—here are some more floating on the surface—wait! No, there aren't any in the healthy stuff where there are yeasts and alcohol. What can that mean?" he pondered. Then he fished down into the bottle and got a speck, with some trouble, into a drop of pure water; he put it before his microscope. . . .

His moment had come.

No yeast globes here, no, but something different, something strange he had never seen before, great tangled dancing masses of tiny rod-like things, some of them alone, some drifting along like strings of boats, all of them shimmying with a weird incessant vibration. He hardly dared to guess at their size—they were much smaller than the yeasts—they were only one-twenty-five-thousandth of an inch long!

That night he tossed and didn't sleep and next morning his

stumpy legs hurried him back to the beet factory. His glasses awry on his nearsighted eyes, he leaned over and dredged up other samples from other sick vats—he forgot all about Bigo and thought nothing of helping Bigo; Bigo didn't exist; nothing in the world existed but his sniffing curious self and these dancing strange rods. In every one of the grayish specks he found millions of them. . . . Feverishly at night with Madame Pasteur waiting up for him and at last going to bed without him, he set up apparatus that made his laboratory look like an alchemist's den. He found that the rod-swarming juice from the sick vats always contained the acid of sour milk—and no alcohol. Suddenly a thought flooded through his brain: "Those little rods in the juice of the sick vats are alive, and it is *they* that make the acid of sour milk—the rods fight with the yeasts perhaps, and get the upper hand. They are the ferment of the sour-milk-acid, just as the yeasts must be the ferment of the alcohol!" He rushed up to tell the patient Madame Pasteur about it, the only half-understanding Madame Pasteur who knew nothing of fermentations, the Madame Pasteur who helped him so by believing always in his wild enthusiasms. . . .

It was only a guess but there was something inside him that whispered to him that it was surely true. There was nothing uncanny about the rightness of his guess; Pasteur made thousands of guesses about the thousand strange events of nature that met his shortsighted peerings. Many of these guesses were wrong—but when he did hit on a right one, how he did test it and prove it and sniff along after it and chase it and throw himself on it and bring it to earth! So it was now, when he was sure he had solved the ten-thousand-year-old mystery of fermentation.

His head buzzed with a hundred confused plans to see if he was really right, but he never neglected the business men and their troubles, or the authorities or the farmers or his students. He turned part of his laboratory into a manure testing station, he hurried to Paris and tried to get himself elected to the Academy of Sciences—and failed—and he took his classes on educational trips to breweries in Valenciennes and foundries in Belgium. In the middle of this he felt sure, one day, that he had a way to prove that the little rods were alive, that in spite of their miserable littleness they did giant's work, the work no giant could do—of changing sugar into lactic acid.

"I can't study these rods that I think are alive in this mixed-up mess of the juice of the beet-pulp from the vats," Pasteur pondered. "I shall have to invent some kind of clear soup for them so that I can see what goes on—I'll have to invent this special food for them and then see if they multiply, if they have young, if a thousand of the small dancing beings appears where there was only one at first." He tried putting some of the grayish specks from the sick vats into pure sugar water. They refused to grow in it. "The rods need a richer food," he meditated, and after many failures he devised a strange soup; he took some dried yeast and boiled it in pure water and strained it so that it was perfectly clear, he added an exact amount of sugar and a little carbonate of chalk to keep the soup from being acid. Then on the point of a fine needle he fished up one of the gray specks from some juice of a sick fermentation. Carefully he sowed this speck in his new clear soup—and put the bottle in an incubating oven—and waited, waited anxious and nervous; it is this business of experiments not coming off at once that is always the curse of microbe hunting.

He waited and signed some vouchers and lectured to students and came back to peer into his incubator at his precious bottle and advised farmers about their crops and fertilizers and bolted absent-minded meals and peered once more at his tubes—and waited. He went to bed without knowing what was happening in his bottle—it is hard to sleep when you do not know such things. . . .

All the next day it was the same, but toward evening when his legs began to be heavy with failure once more, he muttered: "There *is* no clear broth that will let me see these beastly rods growing—but I'll just look once more—"

He held the bottle up to the solitary gaslight that painted grotesque giant shadows of the apparatus on the laboratory walls. "Sure enough, there's something changing here," he whispered; "there are rows of little bubbles coming up from some of the gray specks I sowed in the bottle yesterday—there are many new gray specks—all of them are sprouting bubbles!" Then he became deaf and dumb and blind to the world of men; he stayed entranced before his little incubator; hours floated by, hours that might have been seconds for him. He took up his bottle caressingly; he shook it gently before the light—little spirals of gray murky cloud curled up from the bottom of the flask and from these spirals came big bubbles of gas. Now he would find out!

He put a drop from the bottle before his microscope. Eureka! The field of the lens swarmed and vibrated with shimmying millions of the tiny rods. "They multiply! They are alive!" he whispered to himself, then shouted: "Yes, I'll be up in a little while!" to Madame Pasteur who had called down begging him to come up for dinner, to come for a little rest. For hours he did not come.

Time and again in the days that followed he did the same experiment, putting a tiny drop from a flask that swarmed with rods into a fresh clear flask of yeast soup that had none at all—and every time the rods appeared in billions and each time they made new quantities of the acid of sour milk. Then Pasteur burst out—he was not a patient man—to tell the world. He told Mr. Bigo it was the little rods that made his fermentations sick: "Keep the little rods out of your vats and you'll always get alcohol, Mr. Bigo." He told his classes about his great discovery that such infinitely tiny beasts could make acid of sour milk from sugar—a thing no mere man had ever done or could do. He wrote the news to his old Professor Dumas and to all his friends and he read papers about it to the Lille Scientific Society and sent a learned treatise to the Academy of Sciences in Paris. It is not clear whether Mr. Bigo found it possible to keep the little rods out of his vats—for they were like bad weeds that get into gardens. But to Pasteur that didn't matter so much. Here was the one important fact: *It is living things, sub-visible living beings, that are the real cause of fermentations!*

Innocently he told every one that his discovery was remarkable—he was too much of a child to be modest—and from now on and for years these little ferments filled his sky; he ate and slept and dreamed and loved—after his absent-minded fashion—with his ferments by him. They were his life.

He worked alone for he had no assistant, not even a boy to wash his bottles for him; how then, you will ask, did he find time to cram his days with such a bewildering jumble of events? Partly because he was an energetic man, and partly it was thanks to Madame Pasteur, who in the words of Roux, "loved him even to the point of understanding his work." On those evenings when she wasn't waiting up lonely for him—when she had finished putting to bed those children whose absent-minded father he was—this brave lady sat primly on a straight-backed chair at a little table and wrote scientific papers at his dictation. Again, while he was below brooding over his tubes and bottles she would translate the cramped scrawls of his notebooks into a clear beautiful handwriting. Pasteur was her life and since Pasteur thought only of work her own life melted more and more into his work. . . .

3

Then one day in the midst of all this—they were just nicely settled in Lille—he came to her and said: "We are going to Paris, I have just been made Administrator and Director of Scientific Studies in the Normal School. This is my great chance."

They moved there, and Pasteur found there was absolutely no place for him to work in; there were a few dirty laboratories for the students but none for the professors; what was worse, the Minister of Instruction told him there was not one cent in the budget for those bottles and ovens and microscopes without which he could not live. But Pasteur snooped round in every cranny of the dirty old building and at last climbed tricky stairs to a tiny room where rats played, to an attic under the roof. He chased the rats out and proclaimed this den his laboratory; he got money—in some mysterious way that is still not clear—for his microscopes and tubes and flasks. The world must know how important ferments are in its life. The world soon knew!

His experiment with the little rods that made the acid of sour milk convinced him—why, no one can tell—that other kinds of small beings did a thousand other gigantic and useful and perhaps dangerous things in the world. "It is those yeasts that my microscope showed me in the healthy beet vats, it is those yeasts that turn sugar into alcohol—it is undoubtedly yeasts that make beer from barley and it is certainly yeasts that ferment grapes into wine—I haven't proved it yet, but I know it." Energetically he wiped his fogged spectacles and cheerfully he climbed to his attic. Experiments would tell him; he must make experiments; he must prove to himself he was right— more especially he must prove to the world he was right. But the world of science was against him.

Liebig, the great German, the prince of chemists, the pope of chemistry, was opposed to his idea. "So Liebig says yeasts have nothing to do with the turning of sugar into alcohol— so he claims that you have to have albumen there, and that it is just the albumen breaking down that carries the sugar along down with it, into alcohol." He would show this Liebig! Then a trick to beat Liebig flashed into his head, a crafty trick, a simple clear experiment that would smash Liebig and all other pooh-bahs of chemistry who scorned the important work that his precious microscopic creatures might do.

"What I have to do is to grow yeasts in a soup that has no albumen in it at all. If yeasts will turn sugar into alcohol in such a soup—then Liebig and his theories are finished." Defiance was in every fiber of him. This business was turning from an affair of cold science into a purely personal matter. But it was one thing to have this bright idea and quite another to find an albumenless food for yeasts—yeasts were squeamish in their tastes, confound them—and he fussed around his drafty attic and was for weeks an exasperated, a very grumpy Pasteur. Until one morning a happy accident cleared the road for him.

He had by chance put some salt of ammonia into an albumen soup in which he grew the yeasts for his experiments. "What's this," he meditated. "The ammonia salt keeps disappearing as my yeasts bud and multiply. What does this mean?" He thought, he fumbled— "Wait! The yeasts use up the ammonia salt, they will grow without the albumen!" He slammed shut the door of his attic room, he must be alone while he worked—he loved to be alone as he worked just as he greatly enjoyed spouting his glorious results to worshipful, brilliant audiences. He took clean flasks and poured distilled water into them, and carefully weighed out pure sugar and slid it into this water, and then put in his ammonia salt—it was the tartrate of ammonia that he used. He reached for a bottle that swarmed with young budding yeasts; with care he fished out a yellowish flake of them and dropped it into his new albumenless soup. He put the bottle in his incubating oven. Would they grow?

That night he turned over and over in his bed. He whispered his hopes and fears to Madame Pasteur—she couldn't advise him but she comforted him. She understood everything but couldn't explain away his worries. She was his perfect assistant. . . .

He was back in his attic next morning not knowing how he had got up the stairs, not remembering his breakfast—he might have floated from his bed directly to the rickety dusty incubator that held his flask—that fatal flask. He opened the bottle and put a tiny cloudy drop from it between two thin bits of glass and slid the specimen under the lens of his microscope—and knew the world was his.

"Here they are," he cried, "lovely budding growing young yeasts, hundreds of thousands of them—yes, and here are some of the old ones, the parent yeasts I sowed in the bottle yesterday." He wanted to rush out and tell some one, but he held himself—he must find out something more—he got some of the soup from the fatal bottle into a retort, to find out whether his budding beings had made alcohol. "Liebig is wrong—albumen isn't necessary—it is yeasts, the growth of yeasts that ferments sugar." And he watched trickling tears of alcohol run down the neck of the retort. He spent the next weeks in doing the experiment over and over, to be sure that the yeasts would keep on living, to be certain that they would keep on making alcohol. He transferred them monotonously, from one bottle to another—he put them through countless flasks of this same simple soup of ammonia salt and sugar in water and always the yeasts budded lustily and filled the bottles with a foamy collar of carbonic acid gas. Always they made alcohol! This checking-up of his discoveries was dull work. There was not the excitement, the sleepless waiting for a result he hoped for passionately or feared terribly would not come.

His new fact was old stuff by now but still he kept on, he cared for his yeasts like some tender father, he fed them and loved them and was proud of their miraculous work of turning great quantities of sugar into alcohol. He ruined his health watching them and he violated sacred customs of all good middle-class Frenchmen. He writes of how he sat down before his lens at seven in the evening—and this is the dinner hour of France!—he sat down to watch and see if he could spy on his yeasts in the act of budding. "And from that time," he writes, "I did not take my eye from the microscope." It was half past nine before he was satisfied that he had seen them bud. He made vast crazy tests that lasted from June until September to find out how long yeasts would keep at their work of turning sugar into alcohol, and at the end he cried: "Give your yeasts enough sugar, and they will not stop working for three months, or even more!"

Then for a moment the searcher in him changed into a showman, an exhibitor of stupendous surprises, a missionary in the cause of microbes. The world must know and the people of the world must gasp at this astounding news that millions of gallons of wine in France and boundless oceans of beer in

Germany are not made by men at all but by incessantly toiling
armies of creatures ten billion times smaller than a wee baby!

He read papers about this and gave speeches and threw his
proofs insolently at the great Liebig's head—and in a little
while a storm was up in the little Republic of Science on the
left bank of the Seine in Paris. His old Professors beamed
pride on him and the Academy of Sciences, which had refused
to elect him a member, now gave him the Prize of Physiology,
and the magnificent Claude Bernard—whom Frenchmen
called Physiology itself—praised him in stately sentences. The
next night, Dumas, his old professor—whose brilliant lectures
had made him cry when he was a green boy in Paris—threw
bouquets at Pasteur in a public speech that would have made
another man than Pasteur bow his head and blush and protest.
Pasteur did not blush—he was perfectly sure that Dumas was
right. Instead he sat down proudly and wrote to his father:

"Mr. Dumas, after praising the so great penetration I had
given proof of . . . added: '*The Academy, sir, rewarded you a few
days ago for other profound researches; your audience this evening
will applaud you as one of the most distinguished professors we possess.*' All that I have underlined was said in these very words
by Mr. Dumas, and was followed *by great applause.*"

It is only natural that in the midst of this hurrahing there
was some quiet hissing. Opponents began to rise on all sides.
Pasteur made these enemies not entirely because his discoveries stepped on the toes of old theories and beliefs. No, his
bristling curious impudent air of challenge got him enemies.
He had a way of putting "am-I-not-clever-to-have-found-
this-and-aren't-all-of-you-fools-not-to-believe-it-at-once" between the lines of all of his writings and speeches. He loved
to fight with words, he had a cocky eagerness to get into an
argument with every one about anything. He would have sput-
tered indignantly at an innocently intended comment on his
grammar or his punctuation. Look at portraits of him taken
at this time—it was 1860—read his researches, and you will
find a fighting sureness of his perpetual rightness in every hair

of his eyebrows and even in the technical terms and chemical formulas of his famous scientific papers.

Many people objected to this scornful cockiness—but some good men of science had better reasons for disagreeing with him—his experiments were brilliant, they were startling, but his experiments stopped short of being completely proved. They had loopholes. Every now and then when he set out confidently with some of his gray specks of ferment to make the acid of sour milk, he would find to his disgust a nasty smell of rancid butter wafting up from his bottles. There would be no little rods in the flask—alas—and none of the sour-milk-acid that he had set out to get. These occasional failures, the absence of sure-fire in these tests gave ammunition to his enemies and brought sleepless nights to Pasteur. But not for long! It is not the least strange thing about him that it didn't seem to matter to him that he never quite solved this confusing going wrong of his fermentations; he was a cunning man —instead of butting his head against the wall of this problem, he slipped around it and turned it to his great fame and advantage.

Why this annoying rancid butter smell—why sometimes no sour-milk-acid? One morning, in one of his bottles that had gone bad, he noticed another kind of wee beasts swimming around among a few of the discouraged dancing rods which should have been there in great swarms.

"What are these beasts? They're much bigger than the rods—they don't merely quiver and vibrate—they actually swim around like fish; they must be little animals."

He watched them peevishly, he had an instinct they had no business there. There were processions of them hooked together like barges on the River Seine, strings of clumsy barges that snaked along. Then there were lonely ones that would perform a stately twirl now and again; sometimes they would make a pirouette and balance—the next moment they would shiver at one end in a curious kind of shimmy. It was all very interesting, these various pretty cavortings of these new beasts.

But they had no business there! He tried a hundred ways to keep them out, ways that would seem very clumsy to us now, but just as he thought he had cleaned them out of all his bottles, back they popped. Then one day it flashed over him that every time that his bottles of soup swarmed with this gently moving larger sort of animal, these same bottles of soup had the strong nasty smell of rancid butter.

So he proved, after a fashion, that this new kind of beast was another kind of ferment, a ferment that made the rancid-butter-acid from sugar; but he didn't nail down his proof, because he couldn't be sure, absolutely, that there was one kind and only one kind of beast present in his bottles. While he was a little confused and uncertain about this, he turned his troubles once more to his advantage. He was peering, one day, at the rancid butter microbes swarming before his microscope. "There's something new here—in the middle of the drop they are lively, going every which way." Gently, precisely, a little aimlessly, he moved the specimen so that the edge of the drop was under his lens. . . . "But here at the edge they're not moving, they're lying round stiff as pokers." It was so with every specimen he looked at. "Air kills them," he cried, and was sure he had made a great discovery. A little while afterward he told the Academy proudly that he had not only discovered a new ferment, a wee animal, that had a curious trick of making stale-butter-acid from sugar, but besides this he had discovered that these animals could live and play and move and do their work without any air whatever. Air even killed them! "And this," he cried, "is the first example of little animals living without air!"

Unfortunately it was the third example. Two hundred years before old Leeuwenhoek had seen the same thing. A hundred years later Spallanzani had been amazed to find that microscopic beasts could live without breathing.

Very probably Pasteur didn't know about these discoveries of the old trail blazers—I am sure he was not trying to steal their stuff—but as he went up in his excited climb toward glory and toward always increasing crowds of new discoveries, he regarded less and less what had been done before him and what went on around him. He re-discovered the curious fact that microbes make meat go bad. He failed to give the first discoverer, Schwann, proper credit for it!

But this strange neglect to give credit for the good work of others must not be posted too strongly against him in the Book of St. Peter, because you can see his fine imagination, that poet's thought of his, making its first attempts at showing that microbes are the real murderers of the human race. He dreams in this paper that just as there is putrid meat, so there are putrid diseases. He tells how he suffered in this work with meat gone bad; he tells about the bad smells—and how he hated bad smells!—that filled his little laboratory during these researches: "My researches on the fermentations have led me naturally toward these studies to which I have resolved to devote myself without too much thought of their danger or of the disgust which they inspire in me," and then he told the Academy of the hard job that awaited him; he explained to them why he must not shrink from it, by making a graceful quotation from the great Lavoisier: "Public usefulness and the interests of humanity ennoble the most disgusting work and only allow enlightened men to see the zeal which is needed to overcome obstacles."

4

So he prepared the stage for his dangerous experiments—years before he entered on them. He prepared a public stage-setting. His proposed heroism thrilled the calm men of science that were his audience. As they returned home through the gray streets of the ancient Latin Quarter they could imagine Pasteur bidding them a farewell full of emotion, they could see him marching with set lips—wanting to hold his nose but bravely not doing it—into the midst of stinking pestilences where perilous microbes lay in wait for him. . . . It is so that Pasteur proved himself much more useful than Leeuwenhoek or Spallanzani—he did excellent experiments, and then had a knack of presenting them in a way to heat up the world about them. Grave men of science grew excited. Simple people saw clear visions of the yeasts that made the wine that was their staff of life and they were troubled at nights by thoughts of hovering invisible putrid microbes in the air. . . .

He did curious tests that waited three years to be completed. He took flasks and filled them part way full with milk or urine. He doused them in boiling water and sealed their slender necks shut in a blast flame—then for years he guarded them. At last he opened them, to show that the urine and the milk were perfectly preserved, that the air above the fluid in the bottles still had almost all of its oxygen; no microbes, no destruction of the milk! He allowed germs to grow their silent swarms in other flasks of urine and milk that he had left unboiled, and when he tested these for oxygen he found that the oxygen had been completely used up—the microbes had used it to burn up, to destroy the stuff on which they fed. Then like a great bird Pasteur spread his wings of fancy and soared up to fearsome speculations—he imagined a weird world without microbes, a world whose air had plenty of oxygen, but this oxygen would be of no use, alas, to destroy dead plants and animals, because there were no microbes to do the oxidations. His hearers had nightmare glimpses of vast heaps of carcasses choking deserted lifeless streets—without microbes life would not be possible!

Now Pasteur ran hard up against a question that was bound to pop up and look him in the face sooner or later. It was an old question. Adam had without doubt asked it of God, while he wondered where the ten thousand living beings of the garden of Eden came from. It was the question that had all thinkers by the ears for a hundred centuries, that had given Spallanzani so much exciting fun a hundred years before. It was the simple but absolutely insoluble question: Where do microbes come from?

"How is it," Pasteur's opponents asked him, "how is it that yeasts appear from nowhere every year of every century in every corner of the earth, to turn grape juice into wine?

Where do the little animals come from, these little animals that turn milk sour in every can and butter rancid in every jar, from Greenland to Timbuctoo?"

Like Spallanzani, Pasteur could not believe that the microbes rose from the dead stuff of the milk or butter. Surely microbes have to have parents! He was, you see, a good Catholic. It is true that he lived among the brainy skeptics on the left bank of the Seine in Paris, where God is as popular as a Soviet would be in Wall Street, but the doubts of his colleagues didn't touch Pasteur. It was beginning to be the fashion of the doubters to believe in Evolution: the majestic poem that tells of life, starting as a formless stuff stirring in a steamy ooze of a million years ago, unfolding through a stately procession of living beings until it gets to monkeys and at last— triumphantly—to men. There doesn't have to be a God to start that parade or to run it—it just happened, said the new philosophers with an air of science.

But Pasteur answered: "My philosophy is of the heart and not of the mind, and I give myself up, for instance, to those feelings about eternity that come naturally at the bedside of a cherished child drawing its last breath. At those supreme moments there is something in the depths of our souls which tells us that the world may be more than a mere combination of events due to a machine-like equilibrium brought out of the chaos of the elements simply through the gradual action of the forces of matter." He was always a good Catholic.

Then Pasteur dropped philosophy and set to work. He believed that his yeasts and rods and little animals came from the air—he imagined an air full of these invisible things. Other microbe hunters had shown there were germs in the air, but Pasteur made elaborate machines to prove it all over again. He poked gun cotton into little glass tubes, put a suction pump on one end of them and stuck the other end out of the window, sucked half the air of the garden through the cotton— and then gravely tried to count the number of living beings in this cotton. He invented clumsy machines for getting these microbe-loaded bits of cotton into yeast soup, to see whether the microbes would grow. He did the good old experiment of Spallanzani over; he got himself a round bottle and put some yeast soup in it, and sealed off the neck of the bottle in the stuttering blast lamp flame, then boiled the soup for a few minutes—and no microbes grew in this bottle.

"But you have heated the air in your flask when you boiled the yeast soup—what yeast soup needs to generate little animals is *natural* air—you can't put yeast soup together with natural unheated air without its giving rise to yeasts or molds or torulas or vibrions or animalcules!" cried the believers in spontaneous generation, the evolutionists, the doubting botanists, cried all Godless men from their libraries and their armchairs. They shouted, but made no experiments.

Pasteur, in a muddle, tried to invent ways of getting unheated air into a boiled yeast soup—and yet keep it from swarming with living sub-visible creatures. He fumbled at getting a way to do this; he muddled—keeping all the time a brave face toward the princes and professors and publicists that were now beginning to swarm to watch his miracles. The authorities had promoted him from his rat-infested attic to a little building of four or five two-by-four rooms at the gate of the Normal School. It would not be considered good enough to house the guinea-pigs of the great Institutes of to-day, but it was here that Pasteur set out on his famous adventure to prove that there was nothing to the notion that microbes could arise without parents. It was an adventure that was part good experiment, part unseemly scuffle—a scuffle that threatened at certain hilariously vulgar moments to be settled by a fist fight. He messed around, I say, and his apparatus kept getting more and more complicated, and his experiments kept getting easier to object to and less clear, he began to replace his customary easy experiments that convinced with sledge-hammer force, by long drools of words. He was stuck.

Then one day old Professor Balard walked into his workroom. Balard had started life as a druggist; he had been an owlish original druggist who had amazed the scientific world by making the discovery of the element bromine, not in a fine laboratory, but on the prescription counter in the back room of a drugstore. This had got him fame and his job of professor of chemistry in Paris. Balard was not ambitious; he had no yearning to make all the discoveries in the world—discovering bromine was enough for one man's lifetime—but Balard did like to nose around to watch what went on in other laboratories.

"You say you're stuck, you say you do not see how to get air and boiled yeast soup together without getting living creatures into the yeast soup, my friend?" you can hear the lazy Balard asking the then confused Pasteur. "Look here, you and I both believe there is no such thing as microbes rising in a yeast soup by themselves—we both believe they fall in or creep in with the dust of the air, is it not so?"

"Yes," answered Pasteur, "but—"

"Wait a minute!" interrupted Balard. "Why don't you just try the trick of putting some yeast soup in a bottle, boiling it, then fixing the opening so the dust can't fall in. At the same time the air can get in all it wants to."

"But how?" asked Pasteur.

"Easy," replied the now forgotten Balard. "Take one of your round flasks, put the yeast soup into it, then soften the glass of the flask neck in your blast lamp—and draw the neck out and downward into a thin little tube—turn this little tube down the way a swan bends his neck when he's picking something out of the water. Then just leave the end of the tube open. It's like this—" and Balard sketched a diagram:

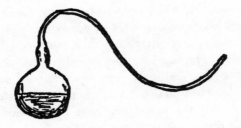

Pasteur looked, then suddenly saw the magnificent ingeniousness of this little experiment. "Why, then microbes can't fall into the flask, because the dust they stick to can't very well fall upward—marvelous! I see it now!"

"Exactly," smiled Balard. "Try it and find out if it works —see you later," and he left to continue his genial round of the laboratories.

Pasteur had bottle washers and assistants now, and he ordered them to hurry and prepare the flasks. In a moment the laboratory was buzzing with the stuttering ear-shattering b-r-r-r-r-r of the enameler's lamps; he fell to work savagely. He took flasks and put yeast soup into them and then melted their necks and drew them out and curved them downward— into swan's necks and pigtails and Chinaman's cues and a half-dozen fantastic shapes. Next he boiled the soup in them—that drove out all the air—but as the flasks cooled down new air came in—unheated air, perfectly clean air.

The flasks ready, Pasteur crawled on his hands and knees, back and forth with a comical dignity on his hands and knees, carrying one flask at a time, through a low cubby-hole under the stairs to his incubating oven. Next morning he was first at the laboratory, and in a jiffy, battered notebook in his hand, if you had been there you would have seen his rear elevation disappearing underneath the stairway. Like a beagle to its rabbit Pasteur was drawn to this oven with its swan neck flasks. Family, love, breakfast, and the rest of a silly world no longer existed for him.

Had you still been there a half hour later you would have seen him come crawling out, his eyes shining through his fogged glasses. He had a right to be happy, for every one of the long twisty necked bottles in which the yeast soup had been boiled was perfectly clear—there was not a living creature in them. The next day they remained the same and the next. There was no doubt now that Balard's scheme had worked. There was no doubt that spontaneous generation was nonsense. "What a fine experiment is this experiment of mine—this proves that you can leave any kind of soup, after you've boiled it, you can leave it open to the ordinary air, and nothing will grow in it—so long as the air gets into it through a narrow twisty tube."

Balard came back and smiled as Pasteur poured the news of the experiment over him. "I thought it would work—you see, when the air comes back in, as the flask cools, the dusts and their germs start in through the narrow neck—but they get caught on the moist walls of the little tube."

"Yes, but how can we prove that?" puzzled Pasteur.

"Just take one of those flasks that has been in your oven all these days, a flask where no living things have appeared, and shake that flask so that the soup sloshes over and back and forth into the swan's neck part of it. Put it back in the oven, and next morning the soup will be cloudy with thick swarms of little beasts—children of the ones that were caught in the neck."

Pasteur tried it, and it was so! A little later at a brilliant meeting where the brains and wit and art of Paris fought to get in, Pasteur told of his swan neck flask experiment in rapturous words. "Never will the doctrine of Spontaneous Generation recover from the mortal blow that this simple experiment has dealt it," he shouted. If Balard was there you may be sure he applauded as enthusiastically as the rest. A rare soul was Balard.

Then Pasteur invented an experiment that was—so far as one can tell from a careful search through the records—really his own. It was a grand experiment, a semi-public experiment, an experiment that meant rushing across France in trains, it was a test in which he had to slither around on glaciers. Once more his laboratory became a shambles of cluttered flasks and hurrying assistants and tinkling glassware and sputtering, bubbling pots of yeast soup. Pasteur and his enthusiastic slaves—they were more like fanatic monks than slaves—were getting ready hundreds of round bellied bottles. They filled each one of them part full of yeast soup and then, during many hours that shot by like moments—such was their excitement—they doused each bottle for a few minutes in boiling water. And while the soup was boiling they drew the flask necks out in a spitting blue flame until they were sealed shut. Each one of this regiment of bottles held boiled yeast soup—and a vacuum.

Armed with these dozens of flasks, and fussing about them, Pasteur started on his travels. He went down first into the dank cellars of the Observatory of Paris, that famous Observatory where worked the great Le Verrier, who had done the proud feat of prophesying the existence of the planet Neptune. "Here the air is so still, so calm," said Pasteur to his boys, "that there will be hardly any dust in it, and almost no microbes." Then, holding the flasks far away from their bodies, using forceps that had been heated red hot in a flame, they cracked the necks of ten of the flasks in succession; as the neck came off each one, there was a hissing "s-s-s-s" of air rushing in. At once they sealed the bottles shut again in the flickering flame of an alcohol lamp. They did the same stunt in the yard of the observatory with another ten bottles, then hurried back to the little laboratory to crawl under the stairs to put the bottles in the incubating oven.

A few days later Pasteur might have been seen squatting before his oven, handling his rows of flasks lovingly, laughing his triumph with one of those extremely rare laughs of his— he only laughed when he found out he was right. He put down tiny scrawls in his notebook, and then crawled out of his cubby-hole to tell his assistants: "Nine out of ten of the bottles we opened in the cellar of the Observatory are perfectly clear—not a single germ got into them. All the bottles we opened in the yard are cloudy—swarming with living creatures. It's the *air* that sucks them into the yeast soup—it's the dust of the air they come in with!"

He gathered up the rest of the bottles and hurried to the train—it was the time of the summer vacation when other professors were resting—and he went to his old home in the Jura mountains and climbed the hill of Poupet and opened twenty bottles there. He went to Switzerland and perilously let the air hiss into twenty flasks on the slopes of Mont Blanc; and found, as he had hoped, that the higher he went, the fewer were the flasks of yeast soup that became cloudy with swarms of microbes. "That is as it ought to be," he cried, "the higher and clearer the air, the less dust—and the fewer the microbes that always stick to particles of the dust." He came back proudly to Paris and told the Academy—with proofs that would astonish everybody!—that it was now sure that air alone could never cause living things to rise in yeast soup. "Here are germs, right beside them there are none, a little further on there are different ones . . . and here where the air is perfectly calm there are none at all," he cried. Then once more he set a new stage for possible magnificent exploits: "I would have liked to have gone up in a balloon to open my bottles still higher up!" But he didn't go up in that balloon, for his hearers were already sufficiently astonished. Already they considered him to be more than a man of science; he became for them a composer of epic searchings, a Ulysses of microbe hunters—the first adventurer of that heroic age to which you will soon come in this story.

Many times Pasteur won his arguments by brilliant experiments that simply floored everyone, but sometimes his victories were due to the weakness or silliness of his opponents, and again they were the result of—luck. Before a society of chemists Pasteur had insulted the scientific ability of naturalists; he was astonished, he shouted, that naturalists didn't stretch out a hand to the real way of doing science—that is, to experiments. "I am of the persuasion that that would put a new sap into their science," he said. You can imagine how the naturalists liked that kind of talk; particularly M. Pouchet, director of the Museum of Rouen, did not like it and he was enthusiastically joined in not liking it by Professor Joly and Mr. Musset, famous naturalists of the College of Toulouse. Nothing could convince these enemies of Pasteur that microscopic beasts did not come to life without parents. They were sure there was such a thing as life arising spontaneously; they decided to beat Pasteur on his own ground at his own game.

Like Pasteur they filled up some flasks, but unlike him they used a soup of hay instead of yeast, they made a vacuum in their bottles and hastened to high Maladetta in the Pyrenees, and they kept climbing until they had got up many feet higher than Pasteur had been on Mont Blanc. Here, beaten upon by nasty breezes that howled out of the caverns of the glaciers and sneaked through the thick linings of their coats, they opened their flasks—Mr. Joly almost slid off the edge of the ledge and was only saved from a scientific martyr's death when a guide grabbed him by the coat tail! Out of breath and chilled through and through they staggered back to a little tavern and put their flasks in an improvised incubating oven—and in a few days, to their joy, they found every one of their bottles swarming with little creatures. Pasteur was wrong!

Now the fight was on. Pasteur became publicly sarcastic about the experiments of Pouchet, Joly and Musset; he made criticisms that to-day we know are quibbles. Pouchet came back with the remark that Pasteur "had presented his own flasks as an ultimatum to science to astonish everybody." Pasteur was furious, denounced Pouchet as a liar and bawled for a public apology. It seemed, alas, as if the truth were going to be decided by the spilling of blood, instead of by calm experiment. Then Pouchet and Joly and Musset challenged Pasteur to a public experiment before the Academy of Sciences, and they said that if one single flask would fail to grow microbes after it had been opened for an instant, they would admit they were wrong. The fatal day for the tests dawned at last—what an interesting day it would have been—but at the last moment Pasteur's enemies backed down. Pasteur did his experiments before the Commission—he did them confidently with ironical remarks—and a little while later the Commission announced: "The facts observed by Mr. Pasteur and contested by Messrs. Pouchet, Joly and Musset, are of the most perfect exactitude."

Luckily for Pasteur, but alas for Truth. Both sides happened to be right. Pouchet and his friends had used hay instead of yeast soup, and a great Englishman, Tyndall, found out years later that hay holds wee stubborn seeds of microbes that will stand boiling for hours! It was really Tyndall that finally settled this great quarrel; it was Tyndall that proved Pasteur was right.

5

Pasteur was now presented to the Emperor Napoleon III. He told that dreamy gentleman that his whole ambition was to find the microbes that he was sure must be the cause of disease. He was invited to an imperial house party at Compiégne. The guests were commanded to get ready to go hunting, but Pasteur begged to be excused; he had had a dray load of apparatus sent up from Paris—though he was only staying at the palace for a week!—and he impressed their Imperial Majesties enormously by bending over his microscope while everybody else was occupied with frivolous and gay amusements.

The world must know that microbes have got to have parents! At Paris he made a popular speech at the scientific soirée at the Sorbonne, before Alexandre Dumas, the novelist, and the woman genius, George Sand, the Princess Mathilde, and a hundred more smart people. That night he staged a scientific vaudeville that sent his audience home in awe and worry; he showed them lantern slides of a dozen different kinds of germs; mysteriously he darkened the hall and suddenly shot a single bright beam of light through the blackness. "Observe the thousands of dancing specks of dust in the path of this ray," he cried; "the air of this hall is filled with these specks of dust, these thousands of little nothings that you should not despise always, for sometimes they carry disease and death; the typhus, the cholera, the yellow fever and many other pestilences!" This was dreadful news; his audience shuddered, convinced by his sincerity. Of course this news was not strictly true, but Pasteur was no mountebank—he believed it himself! Dust and the microbes of the dust had become his life—he was obsessed with dust. At dinner, even at the smartest houses, he would hold his plates and spoons close up to his nose, peer at them, scour them with his napkin, he was with a vengeance putting microbes on the map. . . .

Every Frenchman from the Emperor down was becoming excited about Pasteur and his microbes. Whisperings of mysterious and marvelous events seeped through the gates of the

Normal School. Students, even professors, passed the laboratory a little atremble with awe. One student might be heard remarking to another, as they passed the high gray walls of the Normal School in the Rue d'Ulm: "There is a man working here—his name is Pasteur—who is finding out wonderful things about the machinery of life, he knows even about the origin of life, he is even going to find out, perhaps, what causes disease. . . ." So Pasteur succeeded in getting another year added to the course of scientific studies; new laboratories began to go up; his students shed tears of emotion at the fiery eloquence of his lectures. He talked about microbes causing disease long before he knew anything about whether or not they caused disease—he hadn't yet got his fingers at the throats of mysterious plagues and dreadful deaths, but he knew there were other ways to interest the public, to arouse even such a hardheaded person as the average Frenchman.

"I beg you," he addressed the French people in a passionate pamphlet, "take some interest in those sacred dwellings meaningly called laboratories. Ask that they be multiplied and completed. They are the temples of the future, of riches and comfort." Fifty years ahead of his time as a forward-looking prophet, he held fine austere ideals up to his countrymen while he appealed to their wishes for a somewhat piggish material happiness. A good microbe hunter, he was much more than a mere woolgathering searcher, much more than a mere man of science. . . .

Once more he started out to show all of France how science could save money for her industry; he packed up boxes of glassware and an eager assistant, Duclaux, and bustled off to Arbois, his old home—he hurried off up there to study the diseases of wine—to save the imperiled wine industry. He set up his laboratory in what had been an old café and instead of gas burners he had to be satisfied with an open charcoal brazier that the enthusiastic Duclaux kept glowing with a pair of bellows; from time to time Duclaux would scamper across to the town pump for water; their clumsy apparatus was made by the village carpenter and tinsmith. Pasteur rushed around to his friends of long ago and begged bottles of wine, bitter wine, ropy wine, oily wine; he knew from his old researches that it was yeasts that changed grape juice into wine—he felt certain that it must be some other wee microscopic being that made wines go bad.

Sure enough! When he turned his lens on to ropy wines he found them swarming with very tiny curious microbes hitched together like strings of beads; he found the bottles of bitter wine infested with another kind of beast and the kegs of turned wine by still another. Then he called the winegrowers and the merchants of the region together and proceeded to show them magic.

"Bring me a half dozen bottles of wine that has gone bad with different sicknesses," he asked them. "Do not tell me what is wrong with them, and I'll tell you what ails them without tasting them." The winegrowers didn't believe him; among each other they snickered at him as they went to fetch the bottles of sick wine; they laughed at the fantastic machinery in the old café, they took Pasteur for some kind of earnest lunatic. They planned to fool him and brought him bottles of perfectly good wine among the sick ones. Then he set about flabbergasting them! With a slender glass tube he sucked a drop of wine out of a bottle and put it between two little slips of glass before his microscope. The wine raisers nudged each other and winked French winks of humorous common sense, while Pasteur sat hunched over his microscope, and they became more merry as minutes passed. . . .

Suddenly he looked at them and said: "There is nothing the matter with this wine—give it to the taster—let him see if I'm right."

The taster did his tasting, then puckered up his purple nose and admitted that Pasteur was correct; and so it went through a long row of bottles—when Pasteur looked up from his microscopes and prophesied: "Bitter wine"—it turned out to be bitter; and when he foretold that the next sample was ropy, the taster acknowledged that ropy was right!

The wine raisers mumbled their thanks and lifted their hats

to him as they left. "We don't get the way he does this—but he is a very clever man, very, very clever," they muttered. That is much for a peasant Frenchman to admit. . . .

When they left, Pasteur and Duclaux worked triumphantly in their tumbledown laboratory; they tackled the question of how to keep these microbes out of healthy wines—they found at last that if you heat wine just after it has finished fermenting, even if you heat it gently, way below the point of boiling, the microbes that have no business in the wine will be killed—and the wine will not become sick. That little trick is now known to everybody by the name of pasteurization.

Now that people of the East of France had been shown how to keep their wine from going bad, the people of the middle of France clamored for Pasteur to come and save their vinegar-making industry. So he rushed down to Tours. He had got used to looking for microscopic beings in all kinds of things by now—he no longer groped as he had had to do at first; he approached the vinegar kegs, where wine was turning itself into vinegar, he saw a peculiar-looking scum on the surface of the liquor in the barrels. "That scum has to be there, otherwise we get no vinegar," explained the manufacturers. In a few weeks of swift, sure-fingered investigation that astonished the vinegar-makers and their wives, Pasteur found that the scum on the kegs was nothing more nor less than billions upon billions of microscopic creatures. He took off great sheets of this scum and tested it and weighed it and fussed with it, and at last he told an audience of vinegar-makers and their wives and families that the microbes which change wine to vinegar actually eat up and turn into vinegar ten thousand times their own weight of alcohol in a few days. What gigantic things these infinitely tiny beings can do—think of a man of two hundred pounds chopping two millions of pounds of wood in four days! It was by some such homely comparison as this one that he made microbes part of these humble people's lives, it was so that he made them respect these miserably small creatures; it was by pondering on their fiendish capacity for work that Pasteur himself got used to the idea that there was nothing so strange about a tiny beast, no larger than the microbe of vinegar, getting into an ox or an elephant or a man—and doing him to death. Before he left them he showed the people of Tours how to cultivate and care for those useful wee creatures that so strangely added oxygen to wine to turn it into vinegar—and millions of francs for them.

These successes made Pasteur drunk with confidence in his method of experiment; he began to dream impossible gaudy dreams—of immense discoveries and super-Napoleonic microbe huntings—and he did more than brood alone over these dreams; he put them into speeches and preached them. He became, in a word, a new John the Baptist of the religion of the Germ Theory, but unlike the unlucky Baptist, Pasteur was a forerunner who lived to see at least some of his prophecies come true.

Then for a short time he worked quietly in his laboratory in Paris—there was nothing for him to save just then—until one day in 1865 Fate came to his door and knocked. Fate in the guise of his old professor, Dumas, called on him and asked him to change himself from a man of science into a silkworm doctor. "What's wrong with silkworms? I did not know that they ever had diseases—I know nothing at all about silkworms—what's more, I have never even seen one!" protested Pasteur.

6

"The silk country of the South is my native country," answered Dumas. "I've just come back from there—it is terrible—I cannot sleep nights for thinking of it, my poor country, my village of Alais. . . . This country that used to be rich, that used to be gay with mulberry trees which my people used to call the Golden Tree—this country is desolate now. The lovely terraces are going to ruin—the people, they are my people, they are starving. . . ." Tears were in his voice.

Anything but a respecter of persons, Pasteur who loved and respected himself above all men, had always kept a touching reverence for Dumas. He must help his sad old professor! But how? It is doubtful at this time if Pasteur could have told a silkworm from an angle worm! Indeed, a little later, when he was first given a cocoon to examine, he held it up to his ear, shook it, and cried: "Why, there is something inside it!" Pasteur hated to go South to try to find out what ailed silkworms, he knew he risked a horrid failure by going and he detested failure above everything. But it is one of the charming things about him that in the midst of all his arrogance, his vulgar sureness of himself, he had kept that boyish love and reverence for his old master—so he said to Dumas: "I am in your hands, I'm at your disposal, do with me as you wish—I will go!"

So he went. He packed up the never complaining Madame Pasteur and the children and a microscope and three energetic and worshiping young assistants and he went into the epidemic that was slaughtering millions of silkworms and ruining the South of France. Knowing less of silkworms and their sicknesses than a babe in swaddling clothes he arrived in Alais; he got there and he learned that a silkworm spins a cocoon round itself and turns into a chrysalid inside the cocoon; he found out that the chrysalid changes into a moth that climbs out and lays eggs—which hatch out the next spring into new broods of young silkworms. The silkworm growers—disgusted at his great ignorance—told him that the disease which was killing their worms was called *pébrine*, because the sick worms were covered with little black spots that looked like pepper. Pasteur found out that there were a thousand or so theories about the sickness, but that the little pepper spots—and the curious little globules inside the sick worms, wee globules that you could only see with a microscope—were the only facts that were known about it.

Then Pasteur unlimbered his microscope, before he had got his family settled—he was like one of those trout fishing maniacs who starts to cast without thought of securing his canoe safely on the bank—he unlimbered his microscope, I say, and began to peer at the insides of sick worms, and particularly at these wee globules. Quickly he concluded that the globules were a sure sign of the disease. Fifteen days after he had come to Alais he called the Agricultural Committee together and told them: "At the moment of egg-laying put aside each couple of moths, the father and the mother. Let them mate; let the mother lay her eggs—then pin the father and mother moths down onto a little board, slit open their bellies and take out a little of the fatty tissue under their skin; put this under a microscope and look for those tiny globules. If you can't find any, you can be sure the eggs are sound—you can use those eggs for new silkworms in the spring."

The committee looked at the shining microscope. "We farmers can't run a machine like that," they objected. They were suspicious, they didn't believe in this new-fangled machine. Then the salesman that was in Pasteur came to the front. "Nonsense!" he answered. "There is an eight-year-old girl in my laboratory who handles this microscope easily and is perfectly able to spot these little globules—these corpuscles—and then you grown men try to tell me you couldn't learn to use a microscope!" So he shamed them. And the committee obediently bought microscopes and tried to follow his directions. Then Pasteur started a hectic life; he was everywhere around the tragic silk country, lecturing, asking innumerable questions, teaching the farmers to use microscopes, rushing back to the laboratory to direct his assistants —he directed them to do complicated experiments that he hadn't time to do, or even watch, himself—and in the evenings he dictated answers to letters and scientific papers and speeches to Madame Pasteur. The next morning he was off again to the neighboring towns, cheering up despairing farmers and haranguing them. . . .

But the next spring his bubble burst, alas. The next spring, when it came time for the worms to climb their mulberry twigs to spin their silk cocoons, there was a horrible disaster. His confident prophecy to the farmers did not come true. These honest people glued their eyes to their microscopes to pick out the healthy moths, so as to get healthy eggs, eggs without the evil globules in them—and these supposed healthy eggs hatched worms, sad to tell, who grew miserably, languid worms who would not eat, strange worms who failed to molt, sick worms who shriveled up and died, lazy worms who hung around at the bottoms of their twigs, not caring whether there was ever another silk stocking on the leg of any fine lady in the world.

Poor Pasteur! He had been so busy trying to save the silkworm industry that he hadn't taken time to find out what really ailed the silkworms. Glory had seduced him into becoming a mere savior—for a moment he forgot that Truth is a will o' the wisp that can only be caught in the net of glory-scorning patient experiment. . . .

Some silkworm raisers laughed despairing laughs at him—others attacked him bitterly; dark days were on him. He worked the harder for them, but he couldn't find bottom. He came on broods of silkworms who fairly galloped up the twigs and proceeded to spin elegant cocoons—then at the microscope he found these beasts swarming with the tiny globules. He discovered other broods that sulked on their branches and melted away with a gassy diarrhœa and died miserably—but in these he could find no globules whatever. He became completely mixed up; he began to doubt whether the globules had anything to do with the disease. Then to make things worse, mice got into the broods of his experimental worms and made cheerful meals on them and poor Duclaux, Maillot and Gernez had to stay up by turns all night to catch the raiding mice; next morning everybody would be just started working when black clouds appeared in the West, and all of them—Madame Pasteur and the children bringing up the rear—had to scurry out to cover up the mulberry trees. In the evenings Pasteur had to settle his tired back in an armchair, to dictate answers to peeved silkworm growers who had lost everything—using his method of sorting eggs.

After a series of such weary months, his instinct to do experiments, this instinct—and the Goddess of Chance—came together to save him. He pondered to himself: "I've at least

managed to scrape together a few broods of healthy worms—
if I feed these worms mulberry leaves smeared with the discharges of sick worms, will the healthy worms die?" He tried it,
and the healthy worms died sure enough, but, confound it! the
experiment was a fizzle again—for instead of getting covered
with pepper spots and dying slowly in twenty-five days or so, as
worms always do of *pébrine*—the worms of his experiment
curled up and passed away in seventy-two hours. He was discouraged, he stopped his experiments; his faithful assistants
worried about him—why didn't he try the experiment over?

At last Gernez went off to the north to study the silkworms
of Valenciennes, and Pasteur, not clearly knowing the reason
why, wrote to him and asked him to do that feeding experiment
up there. Gernez had some nice broods of healthy worms. Gernez was sure in his own head—no matter what his chief might
think—that the wee globules were really living things, parasites, assassins of the silkworm. He took forty healthy worms
and fed them on good healthy mulberry leaves that had never
been fed on by sick beasts. These worms proceeded to spin
twenty-seven good cocoons and there were no globules in the
moths that came from them. He smeared some other leaves
with crushed-up sick moths and fed them to some day-old
worms—and these worms wasted away to a slow death, they became covered with pepper spots and their bodies swarmed with
the sub-visible globules. He took some more leaves with
crushed-up sick moths and fed these to some old worms just
ready to spin cocoons; the worms lived to spin the cocoons, but
the moths that came out of the cocoons were loaded with the
globules, and the worms from their eggs came to nothing. Gernez was excited—and he became more excited when still nights
at his microscope showed him that the globules increased tremendously as the worms faded to their deaths. . . .

Gernez hurried to Pasteur. "It is solved," he cried, "the
little globules are alive—they are *parasites!*— They are what
make the worms sick!"

It was six months before Pasteur was convinced that Gernez
was right, but when at last he understood, he swooped back

on his work, and once more called the Committee together. "The little corpuscles are not only a sign of the disease, they are its cause. These globules are alive, they multiply, they force themselves into every part of the moth's body. Where we made our mistake was to examine only a little part of the moth, we only looked under the skin of the moth's belly— we've got to grind up the whole beast and examine all of it. Then if we do not find the globules we can safely use the eggs for next year's worms!"

The committee tried the new scheme and it worked—the next year they had fine worms that gave them splendid yields of silk.

Pasteur saw now that the little globule, the cause of the *pébrine*, came from outside the worm—it did not rise by itself inside the worm—and he went everywhere, showing the farmers how to keep their healthy worms away from all contact with leaves that sick worms had soiled. Then suddenly he fell a victim of a hemorrhage of the brain—he nearly died, but when he heard that work of building his new laboratory had been stopped, frugally stopped in expectation of his death, he was furious and made up his mind to live. He was paralyzed on one side after that—he never got over it—but he earnestly read Dr. Smiles' book, "Self Help" and vigorously decided to work in spite of his handicap. At a time when he should have stayed in his bed, or have gone to the seaside, he staggered to his feet and limped to the train for the South, exclaiming indignantly that it would be criminal not to finish saving the silkworms while so many poor people were starving! All Frenchmen, excepting a few nasty fellows who called it a magnificent gesture, joined in praising him and adoring him.

For six years Pasteur struggled with the diseases of silkworms. He had no sooner settled *pébrine* than another malady of these unhappy beasts popped up, but he knew his problem and found the microbe of this new disease much more quickly. Tears of joy were in the voice of old Dumas now as he thanked his dear Pasteur—and the mayor of the town of Alais talked enthusiastically of raising a golden statue to the great Pasteur.

7

He was forty-five. He wallowed in this glory for a moment, and then—having saved the silkworm industry, with the help of God and Gernez—he raised his eyes toward one of those bright, impossible, but always partly true visions that it was his poet's gift to see. He raised his artist's eyes from the sicknesses of silkworms to the sorrows of men, he sounded a trumpet call of hope to suffering mankind:

"It is in the power of man to make parasitic maladies disappear from the face of the globe, if the doctrine of spontaneous generation is wrong, as I am sure it is."

The siege of Paris in the bitter winter of 1870 had driven him from his work to his old home in the Jura hills. He wandered pitifully around battlefields looking for his son who was a sergeant. Here he worked himself up into a tremendous hate, a hate that never left him, of all things German; he became a professional patriot. "Every one of my works will bear on its title page, 'Hatred to Prussia. Revenge! Revenge!' " he shrieked, good loyal Frenchman that he was. Then with a magnificent silliness he proceeded to make his next research a revenge research. Even he had to admit that French beer was much inferior to the beer of the Germans. Well—he would make the beer of France better than the beer of Germany—he must make the French beer the peer of beers, no, the emperor of all beers of the world!

He embarked on vast voyages to the great breweries of France and here he questioned everybody from the brewmaster in his studio to the lowest workman that cleaned out the vats. He journeyed to England and gave advice to those red-faced artists who made English porter and to the brewers of the divine ale of Bass and Burton. He trained his microscope on the must of a thousand beer vats to watch the yeast globules at their work of budding and making alcohol. Sometimes he discovered the same kind of miserable sub-visible beings that he had found in sick wines years before, and he told the brewers that if they would heat their beer, they would keep these invaders out; he assured them that then they would be able to ship their beer long distances, that then they would be able to brew the most incredibly marvelous of all beers! He begged money for his laboratory from brewers, explaining to them how they would be repaid a thousand fold, and with this money he turned his old laboratory at the Normal School into a small scientific brewery that glittered with handsome copper vats and burnished kettles.

But in the midst of all this feverish work, alas, Pasteur grew sick of working on beer. He hated the taste of beer just as he loathed the smell of tobacco smoke; to his disgust he found that he would have to become a good beer-taster in order to become a great beer-scientist, to his dismay he discovered that there was much more to the art of brewing than simply keeping vicious invading microbes out of beer vats. He puckered his snub nose and buried his serious mustache in foamy mugs and guzzled determined draughts of the product of his pretty kettles—but he detested this beer, even good beer, in fact all beer. Bertin, the physics professor, his old friend, smacked his lips and laughed at him as he swallowed great gulps of beer that Pasteur had denounced as worthless. Even the young assistants snickered—but never to his face. Pasteur, most versatile of men, was after all not a god. He was an investigator and a marvelous missionary—but beer-loving is a gift that is born in a limited number of connoisseurs, just as the ear for telling good music from trash is born in some men!

Pasteur did help the French beer industry. For that we have the testimony of the good brewers themselves. It is my duty to doubt, however, the claims of those idolizers of his who insist that he made French the equal of German beer. I do not deny this claim, but I beg that it be submitted to a commission, one of those solemn impartial international commissions, the kind of commission that Pasteur himself so often demanded to decide before all the world whether he or his detested opponents were in the right. . . .

Pasteur's life was becoming more and more unlike the austere cloistered existence that most men of science lead. His experiments became powerful answers to the objections that swarmed on every side against his theory of germs, they became loud public answers to such objections—rather than calm quests after facts; but in spite of his dragging science into the market place, there is no doubt that his experiments were marvelously made, that they fired the hopes and the imagination of the world. He got himself into a noisy argument on the way yeasts turn grape juice into wine, with two French naturalists, Frémy and Trécul. Frémy admitted that yeasts were needed to make alcohol from grape juice, but he argued ignorantly before the amused Academy that yeasts were spontaneously generated inside of grapes. The wise men of the Academy pooh-poohed; they were amused, all except Pasteur.

"So Frémy says that yeasts rise by themselves inside the grape!" cried Pasteur. "Well, let him answer this experiment then!" He took a great number of round-bellied flasks and filled them part full of grape juice. He drew each one out into a swan's neck; then he boiled the grape juice in all of them for a few minutes and for days and weeks this grape juice, in every one of all these flasks, showed no bubbles, no yeasts, there was no fermentation in them. Then Pasteur went to a vineyard and gathered a few grapes—they were just ripe—and with a pure water he washed the outsides of them with a clean, heated, badger hairbrush. He put a drop of the wash water under his lens—sure enough!—there were globules, a few wee globes, of yeasts. Then he took ten of his swan neck flasks and ingeniously sealed straight tubes of glass into their sides, and through these straight tubes in each one he put a drop of this wash water from the ripe grapes. Presto! Every one of these ten flasks was filled to the neck in a few days with the pink foam of a good fermentation. There was a little of the wash water left; he boiled that and put drops of this through the straight tubes of ten more flasks. "Just so!" he cried a few days later, "there's no fermentation in these flasks, the boiling has killed the yeasts in the wash water."

"Now I shall do the most remarkable experiment of all— I'll prove to this ignorant Frémy that there are no yeasts inside of ripe grapes," and he took a little hollow tube with a sharp point, sealed shut; it was a little tube he had heated very hot in an oven to kill all life—all yeasts—that might have been in it. Carefully he forced the sharp closed point of the tube through the skin into the middle of the grape; delicately he broke the sealed tip *inside* the grape—and the little drop of juice that welled up into the tube he transferred with devilish cunning into another swan-necked flask part filled with grape juice. A few days later he cried, "That finishes Frémy—there is no fermentation in this flask at all—there is no yeast inside the grape!" He went on to one of those sweeping statements he loved to make: "Microbes never rise by themselves inside of grapes, or silkworms, or inside of healthy animals—in animal's blood or urine. All microbes have to get in from the *outside!* That settles Frémy." Then you can fancy him whispering to himself: "The world will soon learn the miracles that will grow from this little experiment."

8

Surely it looked then as if Pasteur had a right to his fantastic dreams of wiping out disease. He had just received a worshiping letter from the English surgeon Lister—and this letter told of a scheme for cutting up sick people in safety, of doing operations in a way that kept out that deadly mysterious infection that in many hospitals killed eight people out of ten. "Permit me," wrote Lister, "to thank you cordially for having shown me the truth of the theory of germs of putrefaction by your brilliant researches, and for having given me the single principle which has made the antiseptic system a success. If you ever come to Edinburgh it will be a real recompense to you, I believe, to see in our hospital in how large a measure humanity has profited from your work."

Like a boy who has just built a steam engine all by himself Pasteur was proud; he showed the letter to all his friends; he inserted it with all its praise in his scientific papers; he published it—of all places—in his book on beer! Then he took a final smash at poor old Frémy, who you would have thought was already sufficiently crushed by the gorgeous experiments; he smashed Frémy not by damning Frémy, but by praising himself! He spoke of his own "remarkable discoveries," he called his own theories the true ones and ended: "In a word, the mark of true theories is their fruitfulness. This is the characteristic which Mr. Balard, with an entirely fatherly friendliness, has made stand out in speaking of my researches." Frémy had no more to say.

All Europe by now was in a furor about microbes, and he knew it was himself that had changed microbes from playthings into useful helpers of mankind—and perhaps, the world would soon be astounded by it—into dread infinitesimal ogres and murdering marauders, the worst enemies of the race. He had become the first citizen of France and even in Denmark prominent brewers were having his bust put in their laboratories. When suddenly Claude Bernard died, and some of Bernard's friends published this great man's unfinished work. Horrible to tell, this unfinished work had for its subject fermentation of grape juice into wine, and it ended by showing that the whole theory of Pasteur was destroyed because . . . and Bernard closed by giving a series of reasons.

Pasteur could not believe his eyes. Bernard had done this, the great Bernard who had been his seatmate in the Academy and had always praised his work; Bernard who had exchanged sly sarcastic remarks with him at the Academy of Medicine about those blue-coated pompous brass-buttoned doctors whose talk was keeping real experiment out of medicine. "It's bad enough for these doctors and these half-witted naturalists to contradict me—but truly great men have always appreciated my work—and now Bernard . . ." you can hear him muttering.

Pasteur was overwhelmed, but only for a moment. He demanded Bernard's original manuscript. They gave it to him. He studied it with all the close attention in his power. He found Bernard's experiments were only beginnings, rough sketches; gleefully he found that Bernard's friends who had published it had made some discreet changes to make it read better. Then he rose one day, to the scandal of the entire Academy and the shocked horror of all the great men of France, and bitterly scolded Bernard's friends for publishing a research that had dared to question his own theories. Vulgarly he shouted objections at Bernard—who, after all, could not answer Pasteur from his grave. Then he published a pamphlet against his old dead friend's last researches. It was a pamphlet in the worst of taste, accusing Bernard of having lost his memory. That pamphlet even claimed that Bernard, who was to his finger tips a hard man of science, had become tainted with mystical ideas by associating too much with literary lights of the French Academy. It even proved that in his last researches Bernard couldn't see well any more—"I'll wager he had become farsighted and could not see the yeasts!" cried Pasteur. Vulgarly, by all this criticism, he left people to conclude that Bernard had been in his dotage when he did his last work—without any sense of the fitness of things this passionate Pasteur jumped up and down on Bernard's grave.

Finally he argued with Bernard by beautiful experiments—a thing most other men would have done without making unseemly remarks. Like an American about to build a skyscraper in six weeks he rushed to carpenters and hardware stores and bought huge pieces of expensive glass and with this glass he had the carpenters build ingenious portable hothouses. His assistants worked dinnerless and sleepless, preparing flasks and microscopes and wads of heated cotton; and in an unbelievably short time Pasteur gathered up all this ponderous paraphernalia and hastened to catch a train for his old home in the Jura mountains. Like the so typical misplaced American that he really was, he threw every consideration and all other work to the winds and went directly to the point of settling: "Does my theory of fermentation hold?"

Coming to his own little vineyard in Arbois, he hastily put up his hothouses around a part of his grape-vines. They were admirable close-fitting hothouses that sealed the grape-vines from the outside air. "It's midsummer, now, the grapes are far from ripe," he pondered, "and I know that at this time there are never any yeasts to be found on the grapes." Then, to make doubly sure that no yeasts from the air could fall on the grapes, he carefully wrapped wads of cotton—which his assistants had heated to kill all living beings—around some of the bunches under the glass of the hothouses. He hurried back to Paris and waited nervously for the grapes to ripen. He went back to Arbois too soon in his frantic eagerness to prove that Bernard was wrong—but at last he got there to find them ripe. He examined the hothouse grapes with his microscope; there was not a yeast to be found on their skins. Feverishly he crushed some of them up in carefully heated bottles—not a single bubble of fermentation rose in these flasks—and when he did the same thing to the exposed grapes from the vines outside the hothouse, these bubbled quickly into wine! At last he gathered up Madame Pasteur and some of the vines with their cotton-wrapped bunches of grapes—he was going to take these back to the Academy, where he would offer a bunch to each member that wanted one, and he was going to challenge everybody to try to make wine from these protected bunches. . . . He knew they couldn't do it without putting yeasts into them. . . . He would show them all Bernard was wrong! Madame Pasteur sat stiffly in the train all the way back to Paris, carefully holding the twigs straight up in front of her so that the cotton wrappings wouldn't come undone. It was a whole day's trip to Paris. . . .

Then at the next meeting Pasteur told the Academy of how he had quarantined his grape-vines against yeasts: "Is it not worthy of attention," he shouted, "that in this vineyard of Arbois, and this would be true of millions of acres of vineyards all over the world, there was at the moment I made these experiments, not a speck of soil which was not capable of fermenting grapes into wine; and is it not remarkable that, on the contrary, the soil of my hothouses could not do this? And why? Because at a definite moment, I covered this soil with some glass. . . ."

Then he jumped to marvelous predictions, prophecies that have since his time come true, he leaped to poetry, I say, that makes you forget his vulgar wrangling with his dead friend Bernard. "Must we not believe, as well, that a day will come when preventive measures that are easy to apply, will arrest those plagues . . ." and he painted them a lurid picture of the terrible yellow fever that just then had changed the gay streets of New Orleans into a desolation. He made them shiver to hear of the black plague on the far banks of the Volga. Finally he made them hope . . .

Meanwhile in a little village in Eastern Germany a young stubborn round-headed Prussian doctor was starting on his road to those very miracles that Pasteur was prophesying— this young doctor was doing strange experiments with mice in time stolen from his practice. He was devising ingenious ways to handle microbes so that he could be dead sure he was handling only one kind—he was learning to do a thing that Pasteur with all his brilliant skill had never succeeded in doing. Let us leave Pasteur for a while—even though he is on the threshold of his most exciting experiments and funniest arguments—let us leave him for a chapter and go with Robert Koch, while he is learning to do fantastic, and marvelously important things with those microbes which had been subjects of Pasteur's kingdom for so many years.

4

Koch

The Death Fighter

1

In those astounding and exciting years between 1860 and 1870, when Pasteur was saving vinegar industries and astonishing emperors and finding out what ailed sick silkworms, a small, serious, and nearsighted German was learning to be a doctor at the University of Göttingen. His name was Robert Koch. He was a good student, but while he hacked at cadavers he dreamed of going tiger-hunting in the jungle. Conscientiously he memorized the names of several hundred bones and muscles, but the fancied moan of the whistles of steamers bound for the East chased this Greek and Latin jargon out of his head.

Koch wanted to be an explorer; or to be a military surgeon and win Iron Crosses; or to be ship's doctor and voyage to impossible places. But alas, when he graduated from the medical college in 1866 he became an interne in a not very interesting insane asylum in Hamburg. Here, busy with raving maniacs and helpless idiots, the echoes of Pasteur's prophecies that there were such things as terrible man-killing microbes hardly reached Koch's ears. He was still listening for steamer-whistles and in the evenings he took walks down by the wharves with Emmy Fraatz; he begged her to marry him; he held out the bait of romantic trips around the world to her. Emmy told Robert that she would marry him, but on condition that he forget this nonsense about an adventurous life, provided that he would settle down to be a practicing doctor, a good useful citizen, in Germany.

Koch listened to Emmy—for a moment the allure of fifty years of bliss with her chased away his dreams of elephants and Patagonia—and he settled down to practice medicine; he began what was to him a totally uninteresting practice of medicine in a succession of unromantic Prussian villages.

Just now, while Koch wrote prescriptions and rode horseback through the mud and waited up nights for Prussian farmer women to have their babies, Lister in Scotland was beginning to save the lives of women in childbirth—by keeping microbes away from them. The professors and the students of the medical colleges of Europe were beginning to be excited and to quarrel about Pasteur's theory of malignant microbes, here and there men were trying crude experiments, but Koch was almost as completely cut off from this world of science as old Leeuwenhoek had been, two hundred years before, when he first fumbled at grinding glass into lenses in Delft in Holland. It looked as if his fate was to be the consoling of sick people and the beneficent and praiseworthy attempt to save the lives of dying people—mostly, of course, he did not save them—and his wife Emmy was quite satisfied with this and was proud when Koch earned five dollars and forty-five cents on especially busy days.

But Robert Koch was restless. He trekked from one deadly village to another still more uninteresting, until at last he came to Wollstein, in East Prussia, and here, on his twenty-eighth birthday, Mrs. Koch bought him a microscope to play with.

You can hear the good woman say: "Maybe that will take Robert's mind off what he calls his stupid practice . . . perhaps this will satisfy him a little . . . he's always looking at everything with his old magnifying glass. . . ."

Alas for her, this new microscope, this plaything, took her husband on more curious adventures than any he would have met in Tahiti or Lahore; and these weird experiences—that Pasteur had dreamed of but which no man had ever had before—came on him out of the dead carcasses of sheep and cows. These new sights and adventures jumped at him impossibly on his very doorstep, and in his own drug-reeking office that he was so tired of, that he was beginning to loathe.

"I hate this bluff that my medical practice is . . . it isn't because I do not *want* to save babies from diphtheria . . . but mothers come to me crying—asking me to save their babies—and what can I do?—Grope . . . fumble . . . reassure them when I know there is no hope. . . . How can I cure diphtheria when I do not even know what causes it, when the wisest doctor in Germany doesn't know? . . ." So you can imagine Koch complaining bitterly to Emmy, who was irritated and puzzled, and thought that it was a young doctor's business to do as well as he could with the great deal of knowledge that he had got at the medical school—oh! would he never be satisfied?

But Koch was right. What, indeed, did doctors know about the mysterious causes of disease? Pasteur's experiments were brilliant, but they had proved nothing about the how and why of human sicknesses. Pasteur was a trail-blazer, a fore-runner crying possible future great victories over disease, shouting about magnificent stampings out of epidemics; but meanwhile the moujiks of desolate towns in Russia were still warding off scourges by hitching four widows to a plow and with them drawing a furrow round their villages in the dead of night—and their doctors had no sounder protection to offer them.

"But the professors, the great doctors in Berlin, Robert, they must know what is the cause of these sicknesses you don't know how to stop." So Frau Koch might have tried to console him. But in 1873—that is only fifty years ago—I must repeat that the most eminent doctors had not one bit better explanation for the causes of epidemics than the ignorant Russian villagers who hitched the town widows to their plows. In Paris

Pasteur was preaching that microbes would soon be found to be the murderers of consumptives: and against this crazy prophet rose the whole corps of the doctors of Paris, headed by the distinguished brass-buttoned Doctor Pidoux.

"What!" roared this Pidoux, "consumption due to a germ —one definite kind of germ? Nonsense! A fatal thought! Consumption is one and many at the same time. Its conclusion is the necrobiotic and infecting destruction of the plasmatic tissue of an organ by a number of roads that the hygienist and the physician must endeavor to close!" It was so that the doctors fought Pasteur's prophecies with utterly meaningless and often idiotic words.

2

Koch was spending his evenings fussing with his new microscope, he was beginning to find out just the right amount of light to shoot up into its lens with the reflecting mirror, he was learning just how needful it was to have his thin glass slides shining clean—those bits of glass on which he liked to put drops of blood from the carcasses of sheep and cows, that had died of anthrax. . . .

Anthrax was a strange disease which was worrying farmers all over Europe, that here and there ruined some prosperous owner of a thousand sheep, that in another place sneaked in and killed the cow—the one support—of a poor widow. There was no rime or reason to the way this plague conducted its maraudings; one day a fat lamb in a flock might be frisking about, that evening this same lamb refused to eat, his head drooped a little—and the next morning the farmer would find him cold and stiff, his blood turned ghastly black. Then the same thing would happen to another lamb, and a sheep, four sheep, six sheep—there was no stopping it. And then the farmer himself, and a shepherd, and a woolsorter, and a dealer in hides might break out in horrible boils—or gasp out their last breaths in a swift pneumonia.

Koch had started using his microscope with the more or

less thorough aimlessness of old Leeuwenhoek; he examined everything under the sun, until he ran on to this blood of sheep and cattle dead of anthrax. Then he began to concentrate, to forget about making a call when he found a dead sheep in a field—he haunted butcher shops to find out about the farms where anthrax was killing the flocks. Koch hadn't the leisure of Leeuwenhoek; he had to snatch moments for his peerings, between prescribing for some child that bawled with a bellyache and the pulling out of a villager's aching tooth. In these interrupted hours he put drops of the blackened blood of a cow dead of anthrax between two thin pieces of glass, very clean shining bits of glass. He looked down the tube of his microscope and among the wee round drifting greenish globules of this blood he saw strange things that looked like little sticks. Sometimes these sticks were short, there might be only a few of them, floating, quivering a little, among the blood globules. But here were others, hooked together without joints—many of them ingeniously glued together till they appeared to him like long threads a thousand times thinner than the finest silk.

"What are these things . . . are they microbes . . . are they alive? They do not move . . . maybe the sick blood of these poor beasts just changes into these threads and rods," Koch pondered. Other men of science, Davaine and Rayer in France, had seen these same things in the blood of dead sheep; and they had announced that these rods were bacilli, living germs, that they were undoubtedly the real cause of anthrax —but they hadn't proved it, and except for Pasteur, no one in Europe believed them. But Koch was not particularly interested in what anybody else thought about the threads and rods in the blood of dead sheep and cattle—the doubts and the laughter of doctors failed to disturb him, and the enthusiasms of Pasteur did not for one moment make him jump at conclusions. Luckily nobody anxious to develop young microbe hunters had ever heard of Koch, he was a lone wolf searcher —he was his own man, alone with the mysterious tangled threads in the blood of the dead beasts.

"I do not see a way yet of finding out whether these little sticks and threads are alive," he meditated, "but there are other things to learn about them. . . ." Then, curiously, he stopped studying diseased creatures and began fussing around with perfectly healthy ones. He went down to the slaughter houses and visited the string butchers and hobnobbed with the meat merchants of Wollstein, and got bits of blood from tens, dozens, fifties of healthy beasts that had been slaughtered for meat. He stole a little more time from his tooth-pullings and professional layings-on-of-hands. More and more Mrs. Koch worried at his not tending to his practice. He bent over his microscope, hours on end, watching the drops of healthy blood.

"Those threads and rods are never found in the blood of any healthy animal," Koch pondered, "—this is all very well, but it doesn't tell me whether they are bacilli, whether they are alive . . . it doesn't show me that they grow, breed, multiply. . . ."

But how to find this out? Consumptives—whom, alas, he could not help—babies choking with diphtheria, old ladies who imagined they were sick, all his cares of a good physician began to be shoved away into one corner of his head. How-to-prove-these-wee-sticks-are-alive, this question made him forget to sign his name to prescriptions, it made him a morose husband, it made him call the carpenter in to put up a partition in his doctor's office. And behind this wall Koch stayed more and more hours, with his microscope and drops of black blood of sheep mysteriously dead—and with a growing number of cages full of scampering white mice.

"I haven't the money to buy sheep and cows for my experiments," you can hear him muttering, while some impatient invalid shuffled her feet in the waiting room, "besides, cows would be a little inconvenient to have around my office—but maybe I can give anthrax to these mice . . . maybe in them I can prove that the sticks really grow. . . ."

So this foiled globe-trotter started on his strange explorations. To me Koch is a still more weird and uncanny microbe hunter than Leeuwenhoek, certainly he was just as much of a self-made scientist. Koch was poor, he had his nose on the grindstone of a medical practice, all the science he knew was what a common medical course had taught him—and from this, God knows, he had learned nothing whatever about the art of doing experiments; he had no apparatus but Emmy's birthday present, that beloved microscope—everything else he had to invent and fashion out of bits of wood and strings and sealing wax. Worst of all, when he came into the living room from his mice and microscope to tell Frau Koch about the new strange things he had discovered, this good lady wrinkled up her nose and told him:

"But, Robert, you smell so!"

Then he hit upon a sure way to give mice the fatal disease of anthrax. He hadn't a convenient syringe with which to shoot the poisonous blood into them, but after sundry cursings and the ruin of a number of perfectly good mice, he took slivers of wood, cleaned them carefully, heated them in an oven to kill any chance ordinary microbes that might be sticking to them. These slivers he dipped into drops of blood from sheep dead of anthrax, blood filled with the mysterious, motionless threads and rods, and then—heaven knows how he managed to hold his wiggling mouse—he made a little cut with a clean knife at the root of the tail of the mouse, and into this cut he delicately slid the blood-soaked splinter. He dropped this mouse into a separate cage and washed his hands and went off in a kind of conscientious wool-gathering way to see what was wrong with a sick baby. . . . "Will that beast, that mouse die of anthrax. . . . Your child will be able to go back to school next week, Frau Schmidt. . . . I hope I didn't get any of that anthrax blood into that cut on my finger . . ." Such was Koch's life.

And next morning Koch came into his home-made laboratory—to find the mouse on its back, stiff, its formerly sleek fur standing on end and its whiteness of yesterday turned into a leaden blue, its legs sticking up in the air. He heated his knives, fastened the poor dead creature onto a board, dissected it, opened it down to its liver and lights, peered into every corner of its carcass. "Yes, this looks like the inside of an anthrax sheep . . . see the spleen, how big, how black it is . . . it almost fills the creature's body. . . ." Swiftly he cut with a clean heated knife into this swollen spleen and put a drop of the blackish ooze from it before his lens. . . .

At last he muttered: "They're here, these sticks and threads . . . they are swarming in the body of this mouse, exactly as they were in the drop of dead sheep's blood that I dipped the little sliver in yesterday." Delighted, Koch knew that he had caused in the mouse, so cheap to buy, so easy to handle, the sickness of sheep and cows and men. Then for a month his life became a monotony of one dead mouse after another, as, day after day, he took a drop of the blood or the spleen of one dead beast, put it carefully on a clean splinter, and slid this sliver into a cut at the root of the tail of a new healthy mouse. Each time, next morning, Koch came into his laboratory to find the new animal had died, of anthrax, and each time in the blood of the dead beast his lens showed him myriads of those sticks and tangled threads—those motionless, twenty-five-thousandth-of-an-inch thick filaments that he could never discover in the blood of any healthy animal.

"These threads *must* be alive," Koch pondered, "the sliver that I put into the mouse has a drop of blood on it and that drop holds only a few hundred of those sticks—and these have grown into billions in the short twenty-four hours in which the beast became sick and died. . . . But, confound it, I must *see* these rods grow—and I can't look inside a live mouse!"

How-shall-I-find-a-way-to-see-the-rods-grow-out-into-threads? This question pounded at him while he counted pulses and looked at his patients' tongues. In the evenings he hurried through supper and growled good-night to Mrs. Koch and shut himself up in his little room that smelled of mice and disinfectant, and tried to find ways to grow his threads outside a mouse's body. At this time Koch knew little or nothing about the yeast soups and flasks of Pasteur, and the experiments he fussed with had the crude originality of the first cave man trying to make fire.

"I will try to make these threads multiply in something that is as near as possible like the stuff an animal's body is made of—it must be just like living stuff," Koch muttered, and he put a wee pin-point piece of spleen from a dead mouse—spleen that was packed with the tangled threads, into a little drop of the watery liquid from the eye of an ox. "That ought to be good food for them," he grumbled. "But maybe, too, the threads have got to have the temperature of a mouse's body to grow," he said, and he built with his own hands a clumsy incubator, heated by an oil lamp. In this uncertain machine he deposited the two flat pieces of glass between which he had put the drop of liquid from the ox-eye. Then, in the middle of the night, after he had gone to bed, but not to sleep, he got up to turn the wick of his smoky incubator lamp down a little, and instead of going back to rest, again and again he slid the thin strips of glass with their imprisoned infinitely little sticks before his microscope. Sometimes he thought he could see them growing—but he could not be sure, because other microbes, swimming and cavorting ones, had an abominable way of getting in between these strips of glass, overgrowing, choking out the slender dangerous rods of anthrax.

"I must grow my rods pure, absolutely pure, without any other microbes around," he muttered. And he kept flounderingly trying ways to do this, and his perplexity pushed up huge wrinkles over the bridge of his nose, and built crow's-feet round his eyes. . . .

Then one day a perfectly easy, a foolishly simple way to watch his rods grow flashed into Koch's head. "I'll put them in a *hanging-drop*, where no other bugs can get in among them," he muttered. On a flat, clear piece of glass, very thin, which he had heated thoroughly to destroy all chance microbes,

Koch placed a drop of the watery fluid of an eye from a just-butchered healthy ox; into this drop he delicately inserted the wee-est fragment of spleen, fresh out of a mouse that had a moment before died miserably of anthrax. Over the drop he put a thick oblong piece of glass with a concave well scooped out of it so that the drop would not be touched. Around this well he had smeared some vaseline to make the thin glass stick to the thick one. Then, dextrously, he turned this simple apparatus upside down, and presto!—here was his hanging-drop, his ox-eye fluid with its rod-swarming spleen, imprisoned in the well—away from all other microbes.

Koch did not know it, perhaps, but this—apart from that day when Leeuwenhoek first saw little animals in rain water—was a most important moment in microbe hunting, and in the fight of mankind against death.

"Nothing can get into that drop—only the rods are there—now we'll see if they will grow," whispered Koch as he slid his hanging-drop under the lens of his microscope; in a kind of stolid excitement he pulled up his chair and sat down to watch what would happen. In the gray circle of the field of his lens he could see only a few shreddy lumps of mouse spleen—they looked microscopically enormous—and here and there a very tiny rod floated among these shreds. He looked—fifty minutes out of each hour for two hours he looked, and nothing happened. But then a weird business began among the shreds of diseased spleen, an unearthly moving picture, a drama that made shivers shoot up and down his back.

The little drifting rods had begun to grow! Here were two where one had been before. There was one slowly stretching itself out into a tangled endless thread, pushing its snaky way across the whole diameter of the field of the lens—in a couple of hours the dead small chunks of spleen were completely hidden by the myriads of rods, the masses of thread that were like a hopelessly tangled ball of colorless yarn, living yarn—silent murderous yarn.

"Now I know that these rods are alive," breathed Koch. "Now I see the way they grow into millions in my poor little mice—in the sheep, in the cows even. One of these rods, these bacilli—he is a billion times smaller than an ox—just one of them maybe gets into an ox, and he doesn't bear any grudge against the ox, he doesn't hate him, but he grows, this bacillus, into millions, everywhere through the big animal, swarming in his lungs and brain, choking his blood-vessels—it is terrible."

Time, his office and its dull duties, his waiting and complaining patients—all of these things became nonsense, seemed of no account, were unreal to Koch whose head was now full of nothing but dreadful pictures of the tangled skeins of the anthrax threads. Then each day of a nervous experiment that lasted eight days Koch repeated his miracle of making a million bacilli grow where only a few were before. He planted a wee bit of his rod-swarming hanging-drop into a fresh, pure drop of the watery fluid of an ox-eye and in every one of these new drops the few rods grew into myriads.

"I have grown these bacilli for eight generations away from any animal, I have grown them pure, apart from any other microbe—there is no part of the dead mouse's spleen, no diseased tissue left in this eighth hanging-drop—only the children of the bacilli that killed the mouse are in it. . . . Will these bacilli still grow in a mouse, or in a sheep, if I inject them—are these threads really the cause of anthrax?"

Carefully Koch smeared a wee bit of his hanging-drop that swarmed with the microbes of the eighth generation—this drop was murky, even to his naked eye, with countless bacilli —he smeared a part of this drop on to a little splinter of wood. Then, with that guardian angel who cares for daring stumbling imprudent searchers of nature standing by him, Koch deftly slid this splinter under the skin of a healthy mouse.

The next day Koch was bending near-sightedly over the body of this little creature pinned on his dissecting board; giddy with hope, he was carefully flaming his knives. . . . Not three minutes later Koch is seated before his microscope, a bit of the dead creature's spleen between two thin bits of glass. "I've proved it," he whispers, "here are the threads, the rods

—those little bacilli from my hanging-drop were just as murderous as the ones right out of the spleen of a dead sheep."

So it was that Koch found in this last mouse exactly the same kind of microbe that he had spied long before—having no idea it was alive—in the blood of the first dead cow he had peered at when his hands were fumbling and his microscope was new. It was precisely the same kind of bacillus that he had nursed so carefully, through long successions of mice, through I do not know how many hanging-drops.

First of all searchers, of all men that ever lived, ahead of the prophet Pasteur who blazed the trail for him, Koch had really made sure that one certain kind of microbe causes one definite kind of disease, that miserably small bacilli may be the assassins of formidable animals. He had angled for these impossibly tiny fish, and spied on them without knowing anything at all of their habits, their lurking places, of how hardy they might be or how vicious, of how easy it might be for them to leap upon him from the perfect ambush their invisibility gave them.

3

Cool and stolid, Koch, now that he had come through these perils, never thought himself a hero; he did not even think of publishing his experiments! To-day it would be inconceivable for a man to do such magnificent work and discover such momentous secrets, and keep his mouth shut about it.

But Koch plugged on, and it is doubtful whether this hesitating, entirely modest genius of a German country doctor realized the beauty or the importance of his lonely experiments.

He plugged on. He must know more! He went pell-mell at the inoculating of guinea-pigs and rabbits, and at last even sheep, with the innocent looking but fatal fluid from the hanging-drops; and in each one of these beasts, in the sheep just as quickly and horribly as the mouse, the few thousands of microbes on the splinter multiplied into billions in the animals, in a few hours they teemed poisonously in what had been robust tissues, choking the little veins and arteries with their myriads, turning to a sinister black the red blood—so killing the sheep, the guinea-pigs, and the rabbits.

At one fantastic jump Koch had soared out of the vast anonymous rank and file of pill-rollers and landed among the most original of the searchers, and the more ingeniously he hunted microbes, the more miserably he tended to the important duties of his practice. Babies in far-off farms howled, but he did not come; peasants, with jumping aches in their teeth, waited sullen hours for him—and at last he had to turn over part of his practice to another doctor. Mrs. Koch saw little of him and worried and wished he would not go on his calls smelling of germicides and of his menagerie of animals. But so far as he was concerned his suffering patients and his wife might have been inhabitants of the other side of the moon—for a new mysterious question was worrying at his head, tugging at him, keeping him awake:

How, in nature, do these little weak anthrax bacilli that fade away and die so easily on my slides, how do they get from sick animals to healthy ones?

There were superstitions among the farmers and horse doctors of Europe about this disease, strange beliefs in regard to the mysterious power of this plague that hung always over their flocks and herds like some cruel invisible sword. Why, this disease is too terrible to be caused by such a wretched little creature as a twenty-thousandth-of-an-inch-long bacillus!

"Your little germ may be what kills our herds, all right, Herr Doktor," the cattle men told Koch, "but how is it that our cows or sheep can be all right in one pasture—perfectly healthy, and then, when we take them into another field, with fine grazing in it, they die like flies?"

Koch knew of this troublesome, mysterious fact too. He knew that in Auvergne in France there were green mountains, horrible mountains where no flock of sheep could go without being picked off, one by one, or in dozens and even hundreds by the black disease, anthrax. And in the country of the Beauce there were fertile fields where sheep grew fat—only to die of anthrax. The peasants shivered at night by their fires: "Our fields are cursed," they whispered.

These things bothered Koch—how could his tiny bacilli live over winter, even for years, in the fields and on the mountains? How could they, indeed, when he had smeared a little bacillus-swarming spleen from a dead mouse on a clean slip of glass, and watched the microbes grow dim, break up, and fade from view? And when he put the nourishing watery fluid of ox-eyes on these bits of glass, the bacilli would no longer grow; when he washed the dried blood off and injected it into mice—these little beasts continued to scamper gayly about in their cages. The microbes, which two days before could have killed a heavy cow, were dead!

"What keeps them alive in the fields, then," muttered Koch, "when they die on my clean glasses in two days?"

Then one day he ran on to a curious sight under his microscope—a strange transformation of his microbes that gave him a clew to his question; and Koch sat down on his stool in his eight-by-ten laboratory in East Prussia and solved the mystery of the cursed fields and mountains of France. He had kept a hanging-drop, in its closed glass well, at the temperature of a mouse's body for twenty-four hours. "Ah, this ought to be full of nice long threads of bacilli," he muttered, and looked down the tube of his microscope—"What's this?" he cried.

The outlines of the threads had grown dim, and each thread was speckled, through its whole length, with little ovals that shone brightly like infinitely tiny glass beads, and these beads were arranged along the threads as perfectly as a string of pearls.

To himself Koch muttered guttural curses. "Other microbes have doubtless gotten into my hanging-drop," he grumbled, but when he looked very carefully he saw that wasn't true, for the shiny little beads were *inside* the threads —the bacilli that made up the threads have turned into these beads! He dried this hanging-drop, and put it away carefully, for a month or so, and then as luck would have it, looked at it once more through his lens. The strange strings of beads were still there, shining as brightly as ever. Then an idea for an experiment got hold of him—he took a drop of pure fresh watery fluid from the eye of an ox. He placed it on the dried-up smear with its months-old bacilli that had turned into beads. His head swam with confused surprise as he looked, and watched the beads grow back into the ordinary bacilli, and then into long threads once more. It was outlandish!

"Those queer shiny beads have turned back into ordinary anthrax bacilli again," cried Koch, "the beads must be the *spores* of the microbe—the tough form of them that can stand great heat, and cold, and drying. . . . That must be the way the anthrax microbe can keep itself alive in the fields for so long—the bacilli must turn into spores. . . ."

Then Koch launched himself into thorough, ingenious tests to see if his quick guess was right. Expertly now he took spleens out of mice which had perished of anthrax—he lifted this deadly stuff out carefully with heated knives and forceps. Protected from all chance of contamination by stray microbes of the air, he kept the spleens for a day at the temperature of a mouse's body, and, sure enough, the microbes, every thread of them, turned into glassy spores.

Then in experiments that kept him incessantly in his dirty little room he found that the spores remained alive for months, ready to hatch out into deadly bacilli the moment he put them into a fresh drop of the watery fluid of ox-eyes, or the instant he stuck them, on one of his thin slivers, into the root of a mouse's tail.

"These spores never form in an animal while he is still alive—they only appear after he has died, and then only when he is kept very warm," said Koch, and he proved this beautifully by clapping spleens into an ice chest—and in a few days this stuff, smeared on splinters, was no more dangerous than if he had shot so much beefsteak into his mice.

It was now the year 1876, and Koch was thirty-four years old, and at last he emerged out of the bush of Wollstein, to tell the world—stuttering a little—that it was at last proved that microbes were the cause of disease. Koch put on his best suit and his gold-rimmed spectacles and packed up his microscope, a few hanging-drops in their glass cells, swarming with murderous anthrax bacilli; and besides these things he bundled a cage into the train with him, a cage that bounced a little with several dozen healthy white mice. He took a train for Breslau to exhibit his anthrax microbes and the way they kill mice, and the weird way in which they turn into glassy spores—he wanted to demonstrate these things to old Professor Cohn, the botanist at the University, who had sometimes written him encouraging letters.

Professor Cohn, who had been amazed at the marvelous experiments about which the lonely Koch had written him, old Cohn snickered when he thought of how this greenhorn doctor—who had no idea, himself, of how original he was—would surprise the highbrows of the University. He sent out invitations to the most eminent medicoes of the school to come to the first night of Koch's show.

4

And they came. To hear the unscientific backwoodsman—they came. They came maybe out of friendliness to old Professor Cohn. But Koch didn't lecture—he was never much at talking—instead of *telling* them that his microbes were the true cause of anthrax, he *showed* these sophisticated professors. For three days and nights he showed them, taking them in swift steps through those searchings he had sweated at—groping and failing often—for years. Never was there a greater comedown for big-wigs who had arrived prepared to be indulgent to a nobody. Koch never argued once, he never bubbled and raved and made prophecies—but he slipped slivers into mouse tails with an unearthly cleverness, and the experienced professors of pathology opened their eyes to see him handle his spores and bacilli and microscopes like a sixty-year-old master. It was a knock-out!

At last Professor Cohnheim, one of the most skillful scientists in the study of diseases in all of Europe, could hold himself no longer. He rushed from the hall, hurried to his own laboratory, and burst into the room where his young student searchers were working. He shouted to them: "My boys, drop everything and go see Doctor Koch—this man has made a great discovery!" Cohnheim gasped to get his breath.

"But who is this Koch, Herr Professor? We've never even heard of him."

"No matter who he is—it is a great discovery, so exact, so simple. It is astounding! This Koch is not a professor, even. . . . He hasn't even been taught how to do research! He's done it all by himself, complete—there is nothing more to do!"

"But what is this discovery, Herr Professor?"

"Go, I tell you, every one of you, and see for yourselves. It is the most marvelous discovery in the realm of microbes . . . he will make us all ashamed of ourselves. . . . Go—" But by this time, all of them, including Paul Ehrlich, had disappeared through the door.

Seven years before, Pasteur had foretold: "It is within the power of man to make parasitic maladies disappear from the face of the earth. . . ." And when he said these words the wisest doctors in the world put their fingers to their heads, thinking: "The poor fellow is cracked!"

But this night Robert Koch had shown the world the first step toward the fulfillment of Pasteur's seemingly insane vision: "Tissues from animals dead of anthrax, whether they are fresh, or putrid, or dried, or a year old, can only produce anthrax when they contain bacilli or the spores of bacilli. Before this fact all doubt must be laid aside that these bacilli are the cause of anthrax," he told them finally, as if his experiments had not convinced them already. And he ended by telling his amazed audience how to fight this terrible disease—how his experiments showed a way to stamp it out in the end: "All animals that die of anthrax must be destroyed at once after they die—or if they can not be burned, they should be buried deep in the ground, where the earth is so cold that the bacilli cannot turn into the tough, long-lived spores. . . ."

So it was that in these three days at Breslau this Koch put a sword Excalibur into the hands of men, with which to begin the fight against their enemies the microbes, their fight against lurking death; so it was that he began to change the whole business of doctors from a foolish hocus-pocus with pills and leeches into an intelligent fight where science instead of superstition was the weapon.

Koch fell among friends—among honest generous men—at Breslau. Cohn and Cohnheim, instead of trying to steal his stuff (there are no fewer shady fellows in science than in any other human activity), these two professors immediately set up a great whooping for Koch, an applause that echoed over Europe and made Pasteur a bit uneasy for his job as Dean of the Microbe Hunters. These two friends began to bombard the authorities of the Imperial Health Office at Berlin about this unknown that Germany ought to be proud of—they did their best to give Koch a chance to do nothing but chase the microbes of disease, to get away from that dull practice of his.

Left alone, or snubbed at Breslau, he might easily have gone back to Wollstein to his business of telling people to stick out their tongues. In short, men of science have either to be showmen—as were the magnificent Spallanzani and the passionate Pasteur—or they have to have impresarios.

Koch packed up Emmy and his household goods and moved to Breslau and was given a job as city physician at four hundred and fifty dollars a year, and was supposed to eke out his living with the private patients that would undoubtedly flock to be treated by such a brilliant man.

So thought Cohn and Cohnheim. But the doorbell of Koch's little office didn't ring, hardly any one came to ring it, and so Koch learned that it is a great disadvantage for a doctor to be brainy and inquire into the final causes of things. He went back to Wollstein, beaten, and here from 1878 to 1880 he made long jumps ahead in microbe hunting once more—spying on and tracking down the strange sub-visible beings that cause the deadly infections of wounds in animals and in human beings. He learned to stain all kinds of bacilli with different colored dyes, so that the very tiniest microbe would stand out clearly. In some unknown way he saved money enough to buy a camera and stuck its lens against his microscope and learned—no one helping him—how to take pictures of these little creatures.

"You'll never convince the world about these murderous bugs until you can show them photographs," Koch said. "Two men can't look through one microscope at the same time, no two men will ever draw the same picture of a germ—so there'll always be wrangling and confusion. . . . But these photographs can't lie—and ten men can study them, and come to an agreement on them. . . ." So it was that Koch began to try to introduce rime and reason into the baby science of microbe hunting which up till now had been as much a wordy brawl as a quest for knowledge.

Meanwhile his friends at Breslau had not forgotten him and in 1880—it was like some bush-leaguer breaking into the big team—he was told by the government to come to Berlin and be Extraordinary Associate of the Imperial Health Office. Here he was given a fine laboratory and a sudden undreamed-of wealth of apparatus and two assistants and enough money so that he could spend sixteen or eighteen hours of his working day among his stains and tubes and chittering guinea-pigs.

By this time the news of Koch's discoveries had spread to all of the laboratories of Europe and had crossed the ocean and inflamed the doctors of America. The vast exciting Battle of the Germ Theory was on! Every medical man and Professor of Diseases who knew—or thought he knew—the top end from the bottom of a microscope set out to become a microbe hunter. Every week brought glad news of the supposed discovery of some new deadly microbe, surely the assassin of suffering from cancer or typhoid fever or consumption. One enthusiast would shout across continents that he had discovered a kind of pan-germ that caused all diseases from pneumonia to the pip—only to be forgotten for an idiot who might claim that he had proved one disease, let us say consumption, to be the result of the attack of a hundred different species of microbes.

So great was the enthusiasm about germs—and the confusion—that Koch's discoveries were in danger of being laughed into obscurity along with the vast magazines full of balderdash that were being printed on the subject of the germ theory.

And yet to-day we demand with a great hue and cry more laboratories, more microbe hunters, better paid searchers to free us from the diseases that scourge us. How futile! For progress, God must send us a few more infernal marvelous searchers of the kind of Robert Koch.

But in the midst of the danger that foolish enthusiasm would kill the new science of microbe hunting, Koch kept his head, and sat down to find a way to grow germs pure. "One germ, one kind of germ only, causes one definite kind of disease—every disease has its own specific microbe, I *know* that," said Koch—without knowing it. "I've got to find a sure easy method of growing one species of germ away from all other contaminating ones that are always threatening to sneak in!"

But how to cage one kind of microbe? All manner of weird machines were being invented to try to keep different sorts of germs apart. Several microbe hunters devised apparatus so complicated that when they had finished building it they probably had already forgotten what they set out to invent it for. To keep stray germs of the air from falling into their bottles some heroic searchers did their inoculations in an actual rain of poisonous germicides!

5

Until, one day, Koch—who frankly admitted it was by accident—looked at the flat surface of half of a boiled potato left on a table in his laboratory. "What's this, I wonder?" he muttered, as he stared at a curious collection of little colored droplets scattered on the surface of the potato. "Here's a gray colored drop, here's a red one, there's a yellow, a violet one —these little specks must be made up of germs from the air. I'll have a look at them."

He stuck his short-sighted eyes down close to the potato so that his scraggly little beard almost dragged in it; he got ready his thin plates of glass and polished off the lenses of his microscope.

With a slender wire of platinum he fished delicately into one of the gray droplets and put a bit of its slimy stuff in a little pure water between two bits of glass, under his microscope. Here he saw a swarm of bacilli, swimming gently about, and every one of these microbes looked exactly like his thousands of brothers in this drop. Then Koch peered at the bugs from a yellow droplet on the potato, and at those of a red one and a violet one. The germs from one were round, from another they had the appearance of swimming sticks, from a third microbes looked like living corkscrews—but all the microbes in one given drop were like their brothers, invariably!

Then in a flash Koch saw the beautiful experiment nature had done for him. "Every one of these droplets is a pure culture of one definite kind of microbe—a pure *colony* of one species of germs. . . . How simple! When germs fall from the air into the liquid soups we have been using—the different kinds of them get all mixed up and swim among each other. . . . But when different bugs fall from the air on the solid surface of this potato—each one has to stay where it falls . . . it sticks there . . . then it grows there, multiplies into millions of its own kind . . . absolutely pure!"

Koch called Loeffler and Gaffky, his two military doctor assistants, and soberly he showed them the change in the whole mixed-up business of microbe hunting that his chance glance at an abandoned potato had brought. It was revolutionary! The three of them set to work with an amazing— loyal Frenchmen might call it stupid—German thoroughness to see if Koch was right. There they sat before the three windows of their room, Koch before his microscope on a high stool in the middle, Loeffler and Gaffky on stools on his left hand and his right—a kind of grimly toiling trinity. They tried to defeat their hopes, but quickly they discovered that Koch's prophecy was an even more true one than he had dreamed. They made mixtures of two or three kinds of germs, mixtures that could never have been untangled by growing in flasks of soup; they streaked these confused species of microbes on the cut flat surfaces of boiled potatoes. And where each separate tiny microbe landed, there it stuck, and grew into a colony of millions of its own kind—and nothing but its own kind.

Now Koch, who, by this simple experience of the old potato, had changed microbe hunting from a guessing game into something that came near the sureness of a science—Koch, I say, got ready to track down the tiny messengers that bring a dozen murderous diseases to mankind. Up till this time Koch had had very little criticism or opposition from other men of science, mainly because he almost never opened his mouth until he was sure of his results. He told of his discoveries with a disarming modesty and his work was so unanswerably complete—he had a way of seeing the objections that critics might make and replying to them in advance—that it was hard to find protestors.

Full of confidence Koch went to Professor Rudolph Virchow, by far the most eminent German researcher in disease, an incredible savant, who knew more than there was to be known about a greater number of subjects than any sixteen scientists together could possibly know. Virchow was, in brief, the ultimate Pooh-Bah of German medical science. He had spoken the very last word on clots in blood vessels and had invented the impressive words, *heteropopia, agenesia,* and *ochronosis,* and many others that I have been trying for years to understand the meaning of. He had—with tremendous mistakenness—maintained that consumption and scrofula were two different diseases; but with his microscope he had made genuinely good, even superb descriptions of the way sick tissues look and he had turned his lens into every noisome nook and cranny of twenty-six thousand dead bodies. Virchow had printed—I do not exaggerate—thousands of scientific papers, on every subject imaginable, from the shapes of little German schoolboys' heads and noses to the remarkably small size of the blood vessels in the bodies of sickly green-faced girls.

Properly awed—as any one would be—Koch tiptoed respectfully into this Presence.

"I have discovered a way to grow microbes pure, unmixed with other germs, Herr Professor," the bashful Koch told Virchow, with deference.

"And how, I beg you tell me, can you do that? It looks to me to be impossible."

"By growing them on solid food—I can get beautiful isolated colonies of one kind of microbe on the surface of a boiled potato. . . . And now I have invented a better way than that . . . I mix gelatin with beef broth . . . and the gelatin sets and makes a solid surface, and—"

But Virchow was not impressed. He made a sardonic remark that it was so hard to keep different races of germs from getting mixed up that Koch would have to have a separate laboratory for each species of microbe. . . . In short, Virchow was very sniffish and cold to Koch, for he had come to that time of life when ageing men believe that everything is known and there is nothing more to be found out. Koch went away a bit depressed, but not one jot was he discouraged. Instead of arguing and writing papers and making speeches against Virchow he launched himself into the most exciting and superb of all his microbe huntings—he set out to spy upon and discover the most vicious of microbes, that mysterious marauder which each year killed one man, woman, and child out of every seven that died, in Europe, in America. Koch rolled up his sleeves and wiped his gold-rimmed glasses and set out to hunt down the microbe of tuberculosis.

6

Compared to this sly murderer the bacillus of anthrax had been reasonably easy to discover—it was a large bug as microbes go, and the bodies of sick animals were literally alive with anthrax germs when the beasts were about to die. But this tubercle germ—if indeed there was such a creature—was a different matter. Many searchers were looking in vain for it. Leeuwenhoek, with his sharpest of all eyes, would never have found it even if he had looked at a hundred sick lungs; Spallanzani's microscopes would not have been good enough to have revealed this sly microbe; Pasteur, searcher that he was, had neither the precise methods of searching, nor, perhaps, the patience, to lay bare this assassin.

All that was known about tuberculosis was that it must be caused by some kind of microbe, since it could be transmitted from sick men to healthy animals. An old Frenchman, Villemin, had pioneered in this work, and Cohnheim, the brilliant professor of Breslau, had found that he could give tuberculosis to rabbits—by putting a bit of the consumptive's sick lung into the front chamber of a rabbit's eye. Here Cohnheim could watch the little islands of sick tissue—the tubercles—spread and do their deadly work; it was a strange clever experiment that was like looking through a window at a disease growing. . . .

Koch had studied Cohnheim's experiments closely. "This is what I need," he meditated. "I may not use human beings for experimental animals, but now I can give the disease, whenever I wish, to animals . . . here is a real chance to study it, handle it, to look for the microbe that must cause it . . . there *must* be a microbe there . . ."

So Koch set to work—he did everything with a cold system that gives one the shivers when one reads his scientific reports—and he got his first consumptive stuff from a powerful man, a laborer aged thirty-six. This man had been superbly healthy three weeks before, when all at once he began to cough, little pains shot through his chest, his body seemed literally to melt away. Four days after this poor fellow entered the hospital, he was dead, riddled with tubercles—every organ was peppered with little grayish-yellow, millet-seed-like specks—

With this dangerous stuff Koch set to work, alone, for Loeffler had set out to track down the microbe of diphtheria and Gaffky was busy trying to find the sub-visible author of typhoid fever. Koch, meanwhile, crushed the yellowish tubercles from the body of the dead man between two heated knives; he ground these granules up and delicately, with a little syringe, injected them into the eyes of numerous rabbits and under the skins of flocks of foolish guinea-pigs. He put these beasts in clean cages and tended them lovingly. And while he waited for his creatures to develop signs of the consumption, he began to peer with his most powerful microscope through the sick tissues that he had taken from the body of the dead workman.

For days he saw nothing. His best lenses, that magnified many hundred times, showed him only the dead ruins of what had once been good healthy lung or liver. "If there is a tubercle microbe, he is such a sneaky fellow that I won't be able, perhaps, to see him in his native state. But I can try painting the tissue with a powerful dye—that may make this bug stand out. . . ."

Day after day, Koch set about staining the stuff from the dead workman brown and blue and violet and most of the colors of the rainbow. Carefully, dipping his hands in the germ-killing bichloride of mercury after almost every move— blackening and wrinkling them with it—he smeared the perilous material from the tubercles on thin clean bits of glass and kept these pieces of glass for hours in a strong blue dye. . . .

Then one morning he took his specimens out of their bath of stain, and put them under his lens, and focussed his microscope and out of the gray mist a strange picture untangled itself. Lying among the shattered diseased lung cells were curious masses of little, infinitely thin bacilli—blue colored rods—so slim that he could not guess their size, and they were less than a fifteen-thousandth of an inch long.

"Ah! they are pretty," he muttered. "They're not straight like the anthrax bugs . . . they have little bends and curves in them. Wait! here are whole bunches of them . . . like cigarettes in a pack—Heh! here is one lone devil *inside* a lung cell . . . I wonder . . . have I found him—that tubercle bug, already?"

Koch went on, precisely, with that efficiency of his, to staining tubercles from every part of the workman's body, and everywhere his blue dye showed up these same slender crooked bacilli—strange creatures unlike any he had seen in all the thousands of animals or men, diseased or healthy, into whose insides he had pried. And now, sorry things began to happen to his inoculated guinea-pigs and rabbits. The guinea-pigs began to huddle disconsolately in the corners of their cages; their sleek coats ruffled and their bouncing little bodies began to fall away until they were sad bags of bones. They were feverish, their cavortings stopped and they looked listlessly at their fine carrots and their fragrant meals of hay—and one by one they died. And as these unconscious martyrs died—for Koch's mad curiosity and for suffering men—the little microbe hunter pinned them down on his post-mortem board and soaked their sick hair with bichloride of mercury and precisely and with breathless care cut them open with sterile knives.

And inside these poor beasts Koch found the same kind of grayish-yellow sinister tubercles that had filled the body of the workman. Into the baths of blue stain on his eternal strips of glass Koch dipped them—and everywhere, in every one, he found the same terrible curved sticks that had jumped into his astounded gaze when he had stained the lung of the dead man.

"I have it!" he whispered, and called the busy Loeffler and the faithful Gaffky from their own spyings on other microbes. "Look!" Koch cried. "One little speck of tubercle I put into this beast six weeks ago—there could not have been more than a few hundred of those bacilli in that small bit—and now they've grown into billions! What devils they are, those germs—from that one place in the guinea-pig's groin they have sneaked everywhere into his body, they have gnawed—they have grown through the walls of his arteries . . . the blood has carried them into his bones . . . into the farthest corner of his brain. . . ."

Now he went to hospitals everywhere in Berlin, and begged the bodies of men or women that had died of consumption, he spent dreary days in dead houses and every evening before his microscope in his laboratory where the stillness was broken only by the eerie purrings and scurryings of guinea-pigs. He injected the sick tissue from the wasted bodies of consumptives who had died, into hundreds of guinea-pigs, into rabbits and three dogs, thirteen scratching cats, ten flopping chickens and twelve pigeons. He didn't stop with these wholesale insane inoculations but shot the same kind of deadly cheesy stuff into white mice and rats and field mice and into two marmots. Never in microbe hunting has there been such appalling thoroughness.

"Ach! this is a little hard on the nerves, this work," he muttered (thinking, perhaps of the lightning move of the paw of one of his cats jabbing the germ-filled syringe needle into his own hand). For Koch, hunting his invisible foes alone, there were so many disagreeable and always imminent possibilities of excitement—of something tragically worse than mere excitement. . . .

But the hand of this completely unheroic looking little microbe hunter never slipped, it just grew drier and more wrinkled and blacker from its incessant baths in the bichloride of mercury—that good bichloride, with which in those old days the groping microbe hunters used to swab down everything, including their own persons. Then, week by week, in all of Koch's meaouwing, crowing, barking, clucking menagerie of beasts those small curved bacilli grew into their relentless millions—and one by one the animals died, and gave eighteen-hour days of work to Robert Koch in post-mortems and blear-eyed peerings through the microscope.

"It is only when a man or beast has tuberculosis that I can find these blue-stained rods, these bacilli," Koch told Loeffler and Gaffky. "In healthy animals—I have looked, you know, at hundreds of them—I never find them."

"That means, without doubt, that you have discovered the bacillus that is the cause, Herr Doktor—"

"No—not yet—what I have done might make Pasteur sure, but I am not at all convinced yet. . . . I have to get these bacilli out of the bodies of my dying animals now . . . grow them on our beef broth jelly, pure colonies of these microbes I must get, and cultivate them for months, away from any living creature . . . and *then*, if I inoculate these cultivations into good healthy animals, and they get tuberculosis . . ." and Koch's sober wrinkled face smiled for a moment. Loeffler and Gaffky, ashamed of their jumping at conclusions, went back awed to their own searchings.

Testing every possible combination that his head could invent, Koch set out to try to grow his bacilli pure on beef-broth jelly. He made a dozen different kinds of good soup for them, he kept his tubes and bottles at the temperature of the room and the temperature of a man's body and the temperature of fever. He cleverly used the sick lungs of guinea-pigs that teemed with bacilli, lungs that held no other stray microbes which might over-grow and choke out those delicate germs which he was sure must be the authors of consumption. The stuff from these lungs he planted dangerously into hundreds of tubes and bottles, but all this work ended in—nothing. In brief, those slim bacilli that grew like weeds in tropic gardens in the bodies of his sick animals, those microbes that swarmed in millions in sick men, those bacilli turned up their noses—that is, they would have if they had been equipped with noses—at the good soups and jellies that Koch cooked for them. It was no go!

But one day a reason for his failures popped into Koch's head: "The trouble is that these tubercle bacilli will only grow in the bodies of living creatures—they are maybe almost *complete* parasites—I must fix a food for them that is as near as possible like the stuff a living animal's body is made of!"

So it was that Koch invented his famous food—blood-serum jelly—for microbes that are too finicky to grow on common provender. He went to string-butchers and got the clear straw-colored serum from the clotted blood of freshly slaughtered healthy cattle and carefully heated this fluid to kill all the stray microbes that might have fallen into it. Delicately he poured this serum into each one of dozens of narrow test-tubes, and placed these on a slant so that there would be a long flat surface on which to smear the sick consumptive tissues. Then ingeniously he heated each tube just hot enough to make the serum set, on a slant, into a clear beautiful jelly.

That morning a guinea-pig, sadly riddled with tuberculosis, had died. He dissected out of it a couple of the grayish yellow tubercles, and then, with a wire of platinum he streaked bits of this bacillus-swarming stuff on the moist surface of his serum jelly, on tube after tube of it. Then, with that drawing in and puffing out of breath that come after a nasty piece of work, well done, Koch took his tubes and put them in the oven—at the exact temperature of a guinea-pig's body.

Day after day Koch hurried in the morning to his incubating oven, and took out his tubes and held them close to his gold-rimmed glasses, and saw—nothing.

"Well, I have failed again," he mumbled—it was the fourteenth day after he had planted his consumptive stuff—"every other microbe I have ever grown multiplies into large colonies in a couple of days, but here, confound it—there is nothing, nothing . . ."

Any other man would have pitched these barren disappointing serum-tubes out, but at this stubbly-haired country doctor's shoulder his familiar demon whispered: "Wait—be patient, my master—you know that tubercle germs sometimes take months, years to kill men. Maybe too they grow very slowly in the serum tubes." So Koch did not pitch the tubes out, and on the morning of the fifteenth day he came back to his incubator—to find the velvety surface of the serum jelly covered with tiny glistening specks! Koch reached a trembling hand for his pocket lens, clapped it to his eye and peered at one tube after another, and through his lens these glistening specks swelled out into dry tiny scales. . . .

In a daze Koch pulled the cotton plug out of one of his tubes, mechanically he flamed its mouth in the sputtering blue fire of the Bunsen burner, with a platinum wire he picked off one of these little flaky colonies—they must be microbes—and not knowing how or what, he got them before his microscope. . . .

Then he knew that he had got to a warm inn on the stony road of his adventure—here they were, countless myriads of these same bacilli, these crooked rods that he had first spied in the lung of the dead workman. They were motionless but surely multiplying and alive—they were delicate and finicky about their food and feeble in size, but more savage than hordes of Huns and more murderous than ten thousand nests of rattlesnakes.

Now Koch, in taut intent months, confirmed his first success—he went after proving it with a patience and a detail that made me sick of his everlasting thoroughness and prudence as I read the endlessly multiplied experiments in his classic report on tuberculosis—from consumptive monkeys and consumptive oxen and consumptive guinea-pigs Koch grew forty-three different families of these deadly rods on his slanted tubes of serum jelly!

And only from animals sick or dying of tuberculosis, could he grow them. For months he nursed these wee murderers along, planting them from one tube to another—with marvelous watchfulness he kept all other chance microbes away from them.

"Now I must shoot these bacilli—these pure cultivations of my bacilli—into healthy guinea-pigs, into all kinds of healthy animals. If then these creatures get tuberculosis, I shall know that my bacilli are necessarily and beyond all doubt the cause!"

That man with the terrible single-mindedness of a maniac driven by a fixed idea changed his laboratory into the weirdest kind of zoo. He became grouchy to every one—to curious visitors he was a sarcastic, spiteful little German ogre. Alone he sterilized batteries of shining syringes and shot the crinkly masses of microbes from the cultivations in his serum-jelly tubes—he injected these bacilli ground up in a little pure water into guinea-pigs and rabbits and hens and rats and mice and monkeys. "That's not enough!" he growled, "I'll try some animals that never are known to have tuberculosis naturally." So he ranged abroad and gathered to his laboratory and injected his beloved terrible bacilli into tortoises, sparrows, five frogs and three eels.

Insanely Koch completed this most fantastic test by sticking his microbes from the serum cultivation into—a goldfish!

Days dragged by, weeks passed, and every day Koch walked into his workshop in the morning and made straight for the cages and jars that held these momentous animals. The goldfish continued to open and shut his mouth and swim placidly about in his round-bellied bowl. The frogs croaked unconcernedly and the eels kept all of their slippery liveliness; the tortoise now and then stuck his head out of his shell and seemed to wink an eye at Koch as if to say: "Your tubercle bugs are food for me—give me some more."

But while his injections worked no harm to these creatures, that do not in the course of nature get consumption anyway —at the same time the guinea-pigs began to droop, to lie pitifully on their sides, gasping. One by one they died, their bodies wasting terribly into tubercles. . . .

Now Koch had forged the last link of the chain of his experiments and was ready to give his news to the world: The bacillus, the true cause of tuberculosis, had been trapped, discovered! When suddenly he decided there was one more thing to do.

"Human beings surely must catch these bacilli by inhaling them, in dust, or from the coughing of people sick with consumption. I wonder, will healthy animals be infected that way too?" At once Koch began to devise ways of doing this experiment—it was a nasty job. "I'll have to *spray* the bacilli from my cultivations at the animals," he pondered. But this was a more serious business than turning ten thousand murderers out of jail. . . .

Like the good hunter that he was, he took a chance with the dangers that he couldn't avoid. He built a big box and put guinea-pigs and mice and rabbits inside it and set this box in the garden. Then through the window he ran a lead pipe that opened in a spray nozzle inside the box, and for three days, for half an hour each day, he sat in his laboratory, pumping at a pair of bellows that shot a poisonous mist of bacilli into the box—to be breathed by the cavorting beasts inside it.

In ten days three of the rabbits were gasping, fighting for that precious air that their sick lungs could no longer give them. In twenty-five days the guinea-pigs had done their humble work—one and all they were dead, of tuberculosis.

Koch told nothing of the ticklish job it was to take these beasts out of their germ-soaked box—if I had been in his place I would rather have handled a boxful of boa-constrictors—and he makes no mention of how he disposed of this little house whose walls had been wet with this so-deadly spray. What chances for making heroic flourishes were missed by this quiet Koch!

7

On the twenty-fourth of March in 1882 in Berlin there was a meeting of the Physiological Society in a plain small room made magnificent by the presence of the most brilliant men of science in Germany. Paul Ehrlich was there and the most eminent Professor Rudolph Virchow—who had but lately sniffed at this crazy Koch and his alleged bacilli of disease—and nearly all of the famous German battlers against disease were there.

A bespectacled wrinkled small man rose and put his face close to his papers and fumbled with them. The papers quivered and his voice shook a little as he started to speak. With an admirable modesty Robert Koch told these men the plain story of the way he had searched out the invisible assassin of one human being out of every seven that died. With no oratorical raisings of his voice he told these disease fighters that the physicians of the world were now able to learn all of the habits of this bacillus of tuberculosis—this smallest but most savage enemy of men. Koch recited to them the lurking places of this slim microbe, its strengths and weaknesses, and he showed them how they might begin the fight to crush, to wipe out this sub-visible deadly enemy.

At last Koch sat down, to wait for the discussion, the inevitable arguments and objections that greet the finish of revolutionary papers. But no man rose to his feet, no word was spoken, and finally eyes began to turn toward Virchow, the oracle, the Tsar of German science, the thunderer whose mere frown had ruined great theories of disease.

All eyes looked at him, but Virchow got up, put on his hat, and left the room—he had no word to say.

If old Leeuwenhoek, two hundred years before, had made so astounding a discovery, Europe of the Seventeenth Century would have heard the news in months. But in 1882 the news that Robert Koch had found the microbe of tuberculosis trickled out of the little room of the Physiological Society the same evening, sang to Kamchatka and to San Francisco on the cable wires that night, and exploded on the front pages of the newspapers in the morning. Then the world went wild over Koch, doctors boarded ships and hopped trains for Berlin to learn from him the secret of hunting microbes; vast crowds of them rushed to Berlin to sit at Koch's feet to learn how to make beef-broth jelly and how to stick syringes full of germs into the wiggling carcasses of guinea-pigs.

Pasteur's deeds had set France by the ears, but Koch's experiments with the dangerous tubercle bacilli rocked the earth, and Koch waved worshipers away, saying:

"This discovery of mine is not such a great advance."

He tried to get away from his adorers and to dodge his eager pupils, to snatch what moments he could for his own new searchings. He loathed teaching—that way he was precisely like Leeuwenhoek—but he was forced, cursing under his breath, to give lessons in microbe hunting to Japanese who spoke horrible German and understood less than they spoke, and to certain Americans, who couldn't by any amount of instruction, learn to hunt microbes. He started a huge fight with Pasteur—but of this I shall tell in the next chapter—and between times he showed his assistant, Gaffky, how to spy on and track down the bacillus of typhoid fever. He was forced to attend idiotic receptions and receive medals, and came away from these occasions to guide his fierce-mustached assistant Loeffler, who was on the trail of the poison-dripping microbe that kills babies with diphtheria. It was thus that Koch shook the tree of his marvelous simple method of growing microbes on the surface of solid food—he shook the tree, as Gaffky said long afterward, and discoveries rained into his lap.

In all of his writings I have never found any evidence that Koch considered himself a great originator; never, like Pasteur, did he seem to realize that he was the leader in the most beautiful and one of the most thrilling battles of men against cruel nature—there was no actor in this mussy-bearded little man. But he did set under way an inspiring drama, a struggle with the messengers of death that turned some of the microbe-hunting actors into maniac searchers, men who went to nearly suicidal lengths, almost murderous extremes—to prove that microbes were the cause of dangerous diseases.

Doctor Fehleisen, to take one instance, went out from Koch's laboratory and found a curious little ball-shaped microbe, hitched to its brothers in chains like the beads of a rosary—he cultivated these bugs from skin gouged out of people sick with erysipelas, that sky-rockety disease that used to be called St. Anthony's Fire. On the theory that an attack of erysipelas might cure cancer—a mad man's excuse!—Fehleisen shot billions of these chain microbes, now known as streptococci, into people hopelessly sick with cancer. And in a few days each one of these human experimental animals of his flamed red with St. Anthony's Fire—some collapsed dangerously and nearly died—and so this desperado proved his case: That streptococcus is the cause of erysipelas.

Another pupil of Koch was the now forgotten hero, Doctor Garré of Basel, who gravely rubbed whole test-tubes full of another kind of microbe—which Pasteur had alleged was the cause of boils—into his own arm. Garré came down horribly with an enormous carbuncle and twenty boils—the tremendous dose of microbes he shot into himself might easily have finished him—but he dismissed his danger as merely "unpleasant" and shouted triumphantly: "I now know that this microbe, this staphylococcus, is the true cause of boils and carbuncles!"

Meanwhile, at the end of 1882, when Koch had finished his virulent and partly comic wrangle with Pasteur, who was just then with prodigious enthusiasm saving the lives of sheep and cattle in France, the discoverer of the tubercle bacillus started sniffing along the trail of one of the most delicate, the most easy to kill, and yet the most terribly savage of all microbes. In 1883 the Asiatic cholera knocked at the door of Europe. This cholera had stolen out of its lurking place in India and slipped mysteriously across the sea and over desert sands to Egypt; suddenly a murderous epidemic of it exploded in Alexandria and Europe across the Mediterranean was frightened. In Alexandria the streets were still with fear; the murderous virus—no one had the slightest notion of what kind of an invisible beast it was—this virus, I say, sneaked into healthy men in the morning, doubled them into knots of spasm-racked agony by afternoon, and put them to rest beyond the reach of all pain by night.

Then a strange race started between Pasteur and Koch, which meant between France and Germany, to search out the microbe of this cholera that flared threatening on the horizon. Koch and Gaffky went armed with microscopes and a menagerie of animals from Berlin; Pasteur—who was desperately busy struggling to conquer the mysterious microbe of hydrophobia—sent the brilliant and devoted Emile Roux and the silent Thuillier, youngest of the microbe hunters of Europe. Koch and Gaffky worked forgetting to eat or sleep; they toiled in dreadful rooms cutting up the bodies of Egyptians dead of cholera; in their muggy laboratory with the air fairly dripping with a steamy heat, sweat dropping off the ends of their noses on to the lenses of their microscopes, they shot stuff from the tragic carcasses of just-dead Alexandrians into apes and dogs and hens and mice and cats. But while these rival teams of searchers hunted frantically the epidemic began to fade away as mysteriously as it came. None of them had yet found a microbe they could surely accuse, and all of them—there is a kind of twisted humor in this—grumbled as they saw death receding, their chance of trapping their prey slipping from them.

Koch and Gaffky were getting ready to return to Berlin, when one morning a frightened messenger came to them and told them: "Dr. Thuillier, of the French Commission, is dead—of cholera."

Koch and Pasteur hated each other sincerely and enthusiastically, like the good patriots that they were, but now the two Germans went to the bereaved Roux and offered their help and their condolences; and Koch was one of those that carried in a plain box to its last home the body of Thuillier, this daring young Thuillier whom the miserably weak—but treacherous—cholera microbe had turned upon and done to death before he had ever had a chance to spy upon and trap it. At the grave Koch laid wreaths upon the coffin: "They are very simple," he said, "but they are of laurel, such as are given to the brave."

The funeral of this first of the martyred microbe hunters over, Koch hurried back to Berlin with certain mysterious boxes that held specimens, that he had painted with powerful dyes, and these specimens had in them a curious microbe shaped like a comma. Koch made his report to the Minister of State: "I have found a germ," he said, "in all cases of cholera . . . but I haven't *proved* yet that it is the cause. Send me to India where cholera is always smoldering—what I have found justifies your sending me there."

So Koch sailed from Berlin for Calcutta, with the fate of Thuillier hanging over him, drolly chaperoning fifty mice and dreadfully annoyed by seasickness. I have often wondered what his fellow-passengers took him for—probably they guessed that he was some earnest little missionary or a serious professor intent to delve into ancient Hindu lore.

Koch found his comma bacillus in the dead bodies of every one of the forty carcasses into which he peered, and he unearthed the same microbe in the intestines of patients at the moment the fatal disease hit them. But he never found this germ in any of the hundreds of healthy Hindus that he examined, nor in any animal, from mice to elephants.

Quickly Koch learned to grow the comma bacillus pure on beef-broth jelly, and once he had it imprisoned in his tubes he studied all the habits of this vicious little vegetable, how it perished quickly when he dried it the least bit, how it could sneak into a healthy man by way of the soiled linen of patients that had died. He dredged this comma microbe up out of the stinking water of the tanks around which clustered the miserable Hindus' huts—sad hovels from which drifted the moans of helpless ones that were dying of cholera.

At last Koch sailed back to Germany, and here he was received like some returning victorious general. "Cholera never rises spontaneously," he told his audience of learned doctors; "no healthy man can ever be attacked by cholera unless he swallows the comma microbe, and this germ can only develop from its like—it cannot be produced from any other thing, or out of nothing. And it is only in the intestine of man, or in highly polluted water like that of India that it can grow."

It is thanks to these bold searchings of Robert Koch that Europe and America no longer dread the devastating raids of these puny but terrible little murderers from the Orient—and their complete extermination from the world waits only upon the civilization and sanitation of India. . . .

8

From the German Emperor's own hand Koch now received the Order of the Crown, with Star, but in spite of that his countrified hat continued to fit his stubbly head, and when admirers adored him he only said to them: "I have worked as hard as I could . . . if my success has been greater than that of most . . . the reason is that I came in my wanderings through the medical field upon regions where the gold was still lying by the wayside . . . and that is no great merit."

The hunters who believed that microbes were the chief foes of man, these men were brave, but there was careless heroism too among some of the ancient doctors and old-fogey sanitarians who thought that all this new stuff about microbes was claptrap and nonsense. Old Professor Pettenkofer of Munich was the leader of the skeptics who were not convinced by Koch's clear experiments, and when Koch came back from India with those comma bacilli that he was sure were the authors of cholera Pettenkofer wrote him something like this: "Send me some of your so-called cholera germs, and I'll show you how harmless they are!"

Koch sent him on a tube that swarmed with wee virulent comma microbes. And so Pettenkofer—to the great alarm of all good microbe hunters—swallowed the entire contents of the tube. There were enough billions of wiggling comma germs in this tube to infect a regiment. Then he growled his scorn through his magnificent beard, and said: "Now let us see if I get cholera!" Mysteriously, nothing happened, and the failure of the mad Pettenkofer to come down with cholera remains to this day an enigma, without even the beginning of an explanation.

Pettenkofer, who was foolhardy enough to try such a possibly suicidal experiment, was also sufficiently cocksure to believe that his drinking of the cholera soup had settled the question in his favor. "Germs are of no account in cholera!" shouted the old doctor. "The important thing is the *disposition* (whatever that means) of the individual!"

"There can be no cholera without the comma bacillus!" said Koch in reply.

"But I have just swallowed millions of your alleged fatal

bacilli, and have not even had a cramp in my stomach!" came back Pettenkofer in rebuttal.

As it is so often the case, alas, in violent scientific controversies, both sides were partly right and partly wrong. Every event of the past forty years has shown that Koch was right when he said that people can never have cholera without swallowing his comma bacillus. And the years that have gone by have revealed that Pettenkofer's experiment pointed out a mystery behind the curtains of the unknown, and these obscuring draperies have not now even begun to be lifted by modern microbe hunters. Murderous germs are everywhere, sneaking into all of us, yet they are able to assassinate only some of us, and that question of the strange resistance of the rest of us is still just as much an unsolved puzzle as it was in those days of the roaring eighteen-eighties when men were ready to risk dying to prove that they were right.

For, make no mistake, Pettenkofer walked within an inch of death; other microbe hunters have since then swallowed cultures of virulent cholera microbes by accident—and died horribly.

But we come to the end of the great days of Robert Koch, and the exploits of Louis Pasteur begin once more to push Koch and all other microbe hunters into the background of the world's attention. Let us leave Koch while his ambitious but well-meaning countrymen prepare, without knowing it, a disaster for him, a tragedy that, alas, has partly tarnished the splendor of his trapping of the microbes that murder animals and men with anthrax and cholera and tuberculosis. But before you read the perfect and brilliant *finale* of the gorgeous career of Pasteur, I beg leave to remove my hat and make bows of respect to Koch—the man who really *proved* that microbes are our most deadly enemies, who brought microbe hunting near to being a science, the man who is now the partly forgotten captain of an obscure heroic age.

5

Pasteur

And the Mad Dog

1

Do not think for a moment that Pasteur allowed his fame and name to be forgotten in the excitement kicked up by the sensational proofs of Koch that microbes murder men. It is certain that less of a hound for sniffing out microbes, less of a poet, less of a master at keeping people wide-eyed with their mouths open, would have been shoved off into a fairly complete oblivion by such events—but not Pasteur!

It was in the late eighteen-seventies—Koch had just swept the German doctors off their feet by his fine discovery of the spores of anthrax—that Pasteur who was only a chemist, had the effrontery to dismiss with a grunt, a shrug, and a wave of his hand, the ten thousand years of experience of doctors in studying and fighting diseases. At this time, in spite of Semmelweis, the Austrian who had proved child-bed fever was contagious, the Lying-In hospitals of Paris were pest-holes. Out of every nineteen women who went hopeful into their doors, one was sure to die of child-bed fever, to leave her baby motherless. One of these places, where ten young mothers perished in succession, was called the House of Crime. Women hardly dared to trust themselves to the most expensive physicians; they were beginning to boycott the hospitals. Large numbers of them—with reason—no longer cared to risk the grim danger of having babies. Even the doctors themselves—accustomed though they were helplessly but sympathetically to preside at the demise of their patients— even the physicians themselves, I say, were scandalized at this dreadful presence of death at the birth of new life.

One day, at the Academy of Medicine in Paris, a famous physician was holding an oration, with plenty of long Greek and elegant Latin words, on the cause—alas, completely unknown to him—of child-bed fever. Suddenly one of his learned and stately sentences was interrupted by a voice bellowing from the rear of the hall:

"The thing that kills women with child-bed fever—it isn't anything like that! It is you doctors that carry deadly microbes from sick women to healthy ones . . . !" It was Pasteur who said this; he was out of his seat; his eyes flamed excitement.

"Possibly you are right, but I fear you will never find that microbe—" The orator tried to start his speech again, but by this time Pasteur was charging up the aisle, dragging his partly paralyzed left leg behind him a little. He reached the blackboard, grabbed a piece of chalk and shouted to the annoyed orator and the scandalized Academy:

"You say I will not find the microbe? Man, I have found it! Here's the way it looks!" And Pasteur scrawled a chain of little circles on the blackboard. The meeting broke up in confusion.

Pasteur was in his late fifties now, but he was still as impetuous and enthusiastic as he had been at twenty-five. He had been a chemist and an expert on beet-sugar fermentations, he had shown the vintners how to keep their wines from spoiling, he had rushed from this job into the saving of sick silkworms, he had preached the slogan of Better Beer for France and had really made the French beer better; but during all these hectic years while he was doing the life work of a dozen men Pasteur dreamed about the tracking down of microbes that he knew must be the scourges of the human race, the authors of disease.

Then suddenly he found Koch had done the trick ahead of him. He must catch up with this Koch. "Microbes are in a way mine—I was the first to show how important they were, twenty years ago, when Koch was a child . . ." you can imagine Pasteur muttering. But there were difficulties in the way of his catching up.

In the first place, Pasteur had never felt a pulse or told a bilious man to stick out his tongue, it is doubtful if he could have told a lung from a liver, and it is certain that he did not know the first thing about how to hold a scalpel. As for those cursed hospitals—phew! The smell of them gave him nasty feelings at the pit of his stomach, and he wanted to stop his ears and run away from the moans that floated down their dingy corridors. But presently—it was ever the way with this unconquerable man—he got around his medical ignorance. Three physicians, Joubert at first, and then Roux and Chamberland became his assistants; youngsters they were, these three, radicals who were Bolshevik against ancient idiotic medical doctrines. They sat worshiping Pasteur at his unpopular lectures in the Academy of Medicine, believing every one of his laughed-at prophecies of dreadful scourges caused by subvisible bugs. He took these boys into his laboratory and in return they explained the machinery of animals' insides to Pasteur, they taught him the difference between the needle and the plunger of a hypodermic syringe and convinced him —he was very squeamish about such things—that animals like guinea-pigs and rabbits hardly felt the prick of the syringe needle when he injected them. Privately these three men swore to be his slaves—and the priests of this new science. . . .

Nothing is truer than that there is no one orthodox way of hunting microbes, and the differences between the ways Koch and Pasteur went at their work are the best illustrations of this. Koch was as coldly logical as a text-book of geometry—he

searched out his bacillus of tuberculosis with systematic experiments, and he thought of all the objections that doubters
might make before such doubters knew that there was any
thing to have doubts about. Koch always recited his failures
with just as much and no more enthusiasm than he did his
triumphs. There was something inhumanly just and right
about him and he looked at his own discoveries as if they had
been those of another man of whom he was a little over-
critical. But Pasteur! This man was a passionate groper whose
head was incessantly inventing right theories and wrong
guesses—shooting them out like a display of village fireworks
going off bewilderingly by accident.

Pasteur started hunting microbes of disease and punched
into a boil on the back of the neck of one of his assistants and
grew a germ from it and was sure it was the cause of boils; he
hurried from these experiments to the hospital to find his
chain microbes in the bodies of women dying with child-bed
fever; from here he rushed out into the country to discover—
but not to prove it precisely—that earthworms carry anthrax
bacilli from the deep buried carcasses of cattle to the surface
of the fields. He was a strange genius who seemed to need the
energetic, gusto-ish doing of a dozen things at the same
time—more or less accurately—in order to discover that grain
of truth which lies at the bottom of most of his work.

In this variety of simultaneous goings-on you can fairly feel
Pasteur fumbling at a way of getting ahead of Koch. Koch had
shown with beautiful clearness that germs cause disease, there
is no doubt about that—but this isn't the most important
thing to do . . . this is nothing, this proof, the thing to do is
to find a way to prevent the germs from killing people, to
protect mankind from death! "What impossible, what absurd
experiments didn't we discuss," said Roux long after this dis-
tressing time when Pasteur was stumbling about in the dark.
"We would laugh at them ourselves, next day."

To understand Pasteur, it is important to know his wild
stabs and his failures as well as his triumphs. He had not the
precise methods of growing microbes pure—it took the

patience of Koch to devise such things—and one day to his disgust, Pasteur observed that a bottle of boiled urine in which he had planted anthrax bacilli was swarming with unbidden guests, contaminating microbes of the air that had sneaked in. The following morning he observed that there were no anthrax germs left at all; they had been completely choked out by the bacilli from the air.

At once Pasteur jumped to a fine idea: "If the harmless bugs from the air choke out the anthrax bacilli in the bottle, they will do it in the body too! It is a kind of dog-eat-dog!" shouted Pasteur, and at once he put Roux and Chamberland to work on the fantastic experiment of giving guinea-pigs anthrax and then shooting doses of billions of harmless microbes into them—beneficent germs which were to chase the anthrax bacilli round the body and devour them—they were to be like the mongoose which kills cobras. . . .

Pasteur gravely announced: "That there were high hopes for the cure of disease from this experiment," but that is the last you hear of it, for Pasteur was never a man to give the world of science the benefit of studying his failures. But a little later the Academy of Sciences sent him on a queer errand, and on this mission he stumbled across a fact that gave him the first clew to a genuine, a remarkable way of turning savage microbes into friendly ones. It was an outlandish plan he began to devise, to dream about, of turning living microbes of disease against their own kind, so guarding animals and men from invisible deaths. At this time there was a great to-do about a cure for anthrax, invented by the horse doctor, Louvrier, in the Jura mountains in the east of France. Louvrier had cured hundreds of cows who were at death's door, said the influential men of the district: it was time that this treatment received scientific approval.

2

Pasteur arrived there, escorted by his young assistants, and found that this miraculous cure consisted first, in having several farm hands rub the sick cow violently to make her as hot as possible; then long gashes were cut in the poor beast's skin and into these cuts Louvrier poured turpentine; finally the now bellowing and deplorably maltreated cow was covered—excepting her face!—with an inch thick layer of unmentionable stuff soaked in hot vinegar. This ointment was kept on the animal—who now doubtless wished she were dead—by a cloth that covered her entire body.

Pasteur said to Louvrier: "Let us make an experiment. All cows attacked by anthrax do not die, some of them just get better by themselves; there is only one way to find out, Doctor Louvrier, whether or no it is your treatment that saves them."

So four good healthy cows were brought, and Pasteur in the presence of Louvrier and a solemn commission of farmers, shot a powerful dose of virulent anthrax microbes into the shoulder of each one of these beasts: this stuff would have surely killed a sheep, it was enough to do to death a few dozen guinea-pigs. The next day Pasteur and the commission and Louvrier returned, and all the cows had large feverish swellings on their shoulders, their breath came in snorts—they were in a bad way, that was very evident.

"Now, Doctor," said Pasteur, "choose two of these sick cows—we'll call them A and B. Give them your new cure, and we'll leave cows C and D without any treatment at all." So Louvrier assaulted poor A and B with his villainous treatment. The result was a terrible blow to the sincere would-be curer of cows, for one of the cows that Louvrier treated got better—but the other perished; and one of the creatures that had got no treatment at all, died—but the other got better.

"Even this experiment might have tricked us, Doctor," said Pasteur. "If you had given your treatment to cows A and D instead of A and B—we all would have thought you had really found a sovereign remedy for anthrax."

Here were two cows left over from the experiment, beasts that had had a hard siege of anthrax and got better from it: "What shall I do with these two cows?" pondered Pasteur. "Well, I might try shooting a still more savage strain of anthrax bacilli into them—I have one family of anthrax germs in Paris that would give even a rhinoceros a bad night."

So Pasteur sent to Paris for his vicious cultivation, and injected five drops into the shoulders of those two cows that had got better. Then he waited, but nothing happened to the beasts, not even a tiny swelling at the point where he had injected millions of poisonous bacilli; the cows remained perfectly happy!

Then Pasteur jumped to one of his quick conclusions: "Once a cow has anthrax, but gets better from it, all the anthrax microbes in the world cannot give her another attack—she is *immune*." This thought began playing and flitting about in his head and made him wool-gather so that he did not hear questions that Madame Pasteur asked him, nor see obvious things at which his eyes looked directly.

"How to give an animal a *little* attack of anthrax, a safe little attack that won't kill him, but will surely protect him. . . . There must be a way to do that. . . . I must find a way."

So it went with Pasteur for months and he kept saying to Roux and Chamberland: "What mystery is there, like the mystery of the non-recurrence of virulent maladies?" He went about muttering to himself: "We must immunize—we must immunize against microbes. . . ."

Meanwhile Pasteur and his faithful crew were training their microscopes on stuff from men and animals dead of a dozen different diseases; there was a kind of mixed-up fumbling in this work between 1878 and 1880—when one day fate, or God, put a marvelous way to immunize right under Pasteur's lucky nose. (It is hard for me to give you this story exactly straight because all of the various people who have written about Pasteur tell it differently and Pasteur himself in his scientific paper says nothing whatever about this remarkable discovery having been a happy accident.) But here it is, as well as I can do, with certain gaps that I have had to fill in myself.

In 1880, Pasteur was playing with the very tiny microbe that kills chickens with a malady known as chicken cholera. Doctor Peronçito had discovered this microbe, so tiny that it was hardly more than a quivering point before the strongest lens. Pasteur was the first microbe hunter to grow it pure, in a soup that he cooked for it from chicken meat. And after he had watched these dancing points multiply into millions in a few hours, he let fall the smallest part of a drop of this bug-swarming broth onto a crumb of bread—and fed this bread to a chicken. In a few hours the unfortunate beast stopped clucking and refused to eat, her feathers ruffled until she looked like a fluffy ball, and the next day Pasteur came in to find the bird tottering, its eyes shut in a kind of invincible drowsiness that turned quickly into death.

Roux and Chamberland nursed these terrible wee microbes along carefully; day after day they dipped a clean platinum needle into a bottle of chicken broth that teemed with germs and then carefully shook the same still-wet needle into a fresh flask of soup that held no microbe at all—so day after day these transplantations went on—always with new myriads of germs growing from the few that had come in on the moistened needle. The benches of the laboratory became cluttered with abandoned cultures, some of them weeks old. "We'll have to clean this mess up to-morrow," thought Pasteur.

Then the god of good accidents whispered in his ear, and Pasteur said to Roux: "We know the chicken cholera microbes are still alive in this bottle . . . they're several weeks old, it is true . . . but just try shooting a few drops of this old cultivation into some chickens. . . ."

Roux followed these directions and the chickens promptly got sick, turned drowsy, lost their customary lively frivolousness. But next morning, when Pasteur came into the laboratory looking for these birds, to put them on the post-mortem board—he was sure they would be dead—he found them perfectly happy and gay!

"This is strange," pondered Pasteur, "always before this the microbes from our cultivations have killed twenty chickens out of twenty . . ." But the time for his discovery was not yet, and next day, after these strangely recovered chickens had been put in charge of the caretaker, Pasteur and his family and Roux and Chamberland went off on their summer vacations. They forgot about those birds. . . .

But at last one day Pasteur told the laboratory servant: "Bring up some healthy birds, new chickens, and get them ready for inoculation."

"But we only have a couple of unused chickens left, Mr. Pasteur—remember, you used the last ones before you went away—you injected the old cultures into them, and they got sick but didn't die?"

Pasteur made a few appropriate remarks about servants who neglected to keep a good supply of fresh chickens on hand. "Well, all right, bring up what new chickens you have left—and let's have a couple of those used ones too—the ones that had the cholera but got better. . . ."

The squawking birds were brought up. The assistant shot the soup with its myriads of germs into the breast muscles of the chickens—into the new ones, *and into the ones that had got better!* Roux and Chamberland came into the laboratory next morning—Pasteur was always there an hour or so ahead of them—they heard the muffled voice of their master shouting to them from the animal room below stairs:

"Roux, Chamberland, come down here—hurry!"

They found him pacing up and down before the chicken cages. "Look!" said Pasteur. "The new birds we shot yesterday—they're dead all right, as they ought to be. . . . But now see these chickens that recovered after we shot them with the old cultures last month. . . . They got the same murderous dose yesterday—but look at them—they have resisted the virulent dose perfectly . . . they are gay . . . they are eating!"

Roux and Chamberland were puzzled for a moment.

Then Pasteur raved: "But don't you see what this means? Everything is found! Now I have found out how to make a beast a little sick—just a little sick so that he will get better, from a disease. . . . All we have to do is to let our virulent microbes grow old in their bottles . . . instead of planting them into new ones every day. . . . When the microbes age, they get tame . . . they give the chicken the disease . . . but only a little of it . . . and when she gets better she can stand all the vicious virulent microbes in the world. . . . This is our chance—this is my most remarkable discovery—this is a *vaccine* I've discovered, much more sure, more scientific than the one for smallpox where no one has seen the germ. . . . We'll apply this to anthrax too . . . to all virulent diseases. . . . We will save lives . . . !"

3

A lesser man than Pasteur might have done this same accidental experiment—for this was no test planned by the human brain—a lesser man might have done it and would have spent years trying to explain to himself the mystery of it, but Pasteur stumbling on this chance protection of a couple of miserable chickens, saw at once a new way of guarding living things against virulent germs, of saving men from death. His brain jumped to a new way of tricking the hitherto inexorable God who ruled that men must be helpless before the sneaking attacks of his sub-visible enemies. . . .

Pasteur was fifty-eight years old now, he was past his prime, but with this chance discovery of the vaccine that saved chickens from cholera, he started the six most hectic years of his life, years of appalling arguments and unhoped-for triumphs and terrible disappointments—into these years, in short, he poured the energy and the events of the lives of a hundred ordinary men.

Hurriedly Pasteur and Roux and Chamberland set out to confirm the first chance observation they had made. They let virulent chicken cholera microbes grow old in their bottles of broth; they inoculated these enfeebled bugs into dozens of healthy chickens—which promptly got sick, but as quickly recovered. Then triumphantly, a few days later, they watched these birds—these *vaccinated* chickens—tolerate murderous injections of millions of microbes, enough to kill a dozen new birds who were not immune.

So it was that Pasteur, ingeniously, turned microbes against themselves. He tamed them first, and then he strangely used them for wonderful protective weapons against the assaults of their own kind.

And now Pasteur, with his characteristic impetuousness— after all it was only chickens he had learned to guard from death so far—became more arrogant than ever with the old-fashioned doctors who talked Latin words and wrote shot-gun prescriptions. He went to a meeting of the Academy of Medicine and with complaisance told the doctors how his chicken vaccinations were a great advance on the immortal smallpox discovery of Jenner: "In this case I have demonstrated a thing that Jenner never could do in smallpox—and that is, that the microbe that kills is the same one that guards the animal from death!"

The old-fashioned blue-coated doctors were peeved at Pasteur's appointing himself a god superior to the great Jenner; Doctor Jules Guérin, the famous surgeon, became particularly sarcastic about Pasteur making so much of mere fussings with chickens—and the fight was on. Pasteur, in a fury got up and shouted remarks about the utter nonsensicality of one of Guérin's pet operations, and there occurred a most scandalous scene—it embarrasses me to have to tell about it—a strange shambles in which Guérin, who was past eighty, rose from his seat and was about to fall on the sixty-year-old Pasteur. The old man aimed a wallop at Pasteur, but frantic friends jumped in and prevented the impending fisticuffs of these two men who thought they could settle the truth by kicks and blows and mayhem.

Next day the ancient Guérin sent his seconds to Pasteur with a challenge to a duel, but Pasteur, evidently, did not care to risk dying that way and he sent Guérin's friends to the Secretary of the Academy with this message: "I am ready, having no right to act otherwise, to modify whatever the editors may consider as going beyond the rights of criticism and legitimate defense." And so Pasteur once more proved himself to be a human being—if not what is commonly called a man —by backing out of the fight.

As I have told you before, Pasteur had a great deal of the mystic in him. Often he bowed himself down before that mysterious Infinite—he worshiped the Infinite when he was not clutching at it like a baby reaching for the moon; but frequently, the moment one of his beautiful experiments had knocked another little chunk off that surrounding Unknown, he made the mistake of believing that all mysteries had dissolved away. It was so now—when he saw that he could really protect chickens perfectly against a fatal illness by his amazing trick of sticking a few of their own tamed assassins into them. At once Pasteur guessed: "Maybe these fowl-cholera microbes will guard chickens against other virulent diseases!" and promptly he inoculated some hens with his new vaccine of weakened fowl cholera germs and then injected them with some certainly murderous *anthrax* bacilli—and the chickens did not die!

Wildly excited he wrote to Dumas, his old professor, and hinted that the new fowl-cholera vaccine might be a wonderful Pan-Protector against all kinds of virulent maladies. "If this is confirmed," he wrote, "we can hope for the most important consequences, even in human maladies."

Old Dumas, greatly thrilled, had this letter published in the Reports of the Academy of Sciences, and there it stands, a sad monument to Pasteur's impetuousness, a blot on his record of reporting nothing but *facts*. So far as I can find, Pasteur never retracted this error, although he soon found that a vaccine made from one kind of bacillus does not protect an animal against all diseases, but only—and then not absolutely surely—against the one disease of which the microbe in the vaccine is the cause.

But one of Pasteur's most charming traits was his characteristic of a scientific Phœnix, who rose triumphantly from the ashes of his own mistakes. When his imagination carried him into the clouds you find him presently landing on the ground with a bump—making clever experiments again, digging for good true hard facts. So it is not surprising to find him, with Roux and Chamberland, in 1881, discovering a very pretty way of taming vicious anthrax microbes and turning them into a vaccine. By this time the quest after vaccines had become so violent that Roux and Chamberland hardly had their Sundays off, and never went on vacations; they slept at the laboratory to be near their tubes and microscopes and microbes. And here, Pasteur directing them, they delicately weakened anthrax bacilli so that some killed guinea-pigs, but not rabbits, and others did mice to death, but were too weak to harm guinea-pigs. They shot the weaker and then the stronger microbes into sheep, who got a little sick but then recovered, and after that these sheep could stand, apparently, the assaults of vicious anthrax germs that were able to kill even a cow.

At once Pasteur told this new triumph to the Academy of Sciences—he had left off going to the Academy of Medicine after his brawl with Guérin—and he held out purple hopes to them that he would presently invent ingenious vaccines that would wipe out all diseases from mumps to malaria. "What is more easy," he shouted, "than to find in these successive viruses a vaccine capable of making sheep and cows and horses a little sick with anthrax without letting them perish—and so preserving them from subsequent maladies?" Some of Pasteur's colleagues thought he was a little cocksure about this, and they ventured to protest. Pasteur's veins stood out on his forehead, but he managed to keep his mouth shut until he and Roux were on the way home, when he burst out, speaking really of all people who failed to see the absolute truth of his idea:

"I would not be surprised if such a man were to be caught beating his wife!"

Make no mistake—science was no cool collecting of facts for Pasteur; in him it set going the same kind of machinery that stirs the human animal to tears at the death of a baby and makes him sing when he hears his uncle has died and left him five hundred thousand dollars.

But enemies were on Pasteur's trail again. Just as he was always stepping on the toes of physicians, so he had offended the high and useful profession of the horse doctors, and one of the leading horse doctors, the editor of one of the most important journals of horse doctoring, his name was Doctor Rossignol, cooked up a plot to lure Pasteur into a dangerous public experiment and so destroy him. This Rossignol got up with a great show of scientific fairness at the Agricultural Society of Melun and said:

"Pasteur claims that nothing is easier than to make a vaccine that will protect sheep and cows absolutely from anthrax. If that is true, it would be a great thing for French farmers, who are now losing twenty million francs a year from this disease. Well, if Pasteur can really make such magic stuff, he ought to be willing to prove to us that he has the goods. Let us get Pasteur to consent to a grand public experiment; if he is right, we farmers and veterinarians are the gainers—if it fails, Pasteur will have to stop his eternal blabbing about great discoveries that save sheep and worms and babies and hippopotamuses!" Like this argued the sly Rossignol.

At once the Society raised a lot of francs to buy forty-eight sheep and two goats and several cows and the distinguished old Baron de la Rochette was sent to flatter Pasteur into this dangerous experiment.

But Pasteur was not one bit suspicious. "Of course I am willing to demonstrate to your society that my vaccine is a life-saver—what will work in the laboratory on fourteen sheep will work on sixty at Melun!"

That was the great thing about Pasteur! When he prepared to take the rabbit out of the hat, to astonish the world, he was absolutely sincere about it; he was a magnificent showman and not below some small occasional hocus-pocus, but he was no designing mountebank. And the public test was set for May and June, that year.

Roux and Chamberland—who had begun to see animals that were strange combinations of chickens and guinea-pigs in their dreams, to drop important flasks, to lie awake injecting millions of imaginary guinea-pigs, these fagged-out boys had just started off on a vacation to the country—when they received telegrams that brought them back to their exciting treadmill:

COME BACK PARIS AT ONCE ABOUT TO MAKE PUBLIC DEMONSTRATION THAT OUR VACCINE WILL PROTECT SHEEP AGAINST ANTHRAX—L. PASTEUR.

Something like that read these wires.

They hurried back. Pasteur said to them: "Before the Agricultural Society of Melun, at the farm of Pouilly-le-Fort, I am going to vaccinate twenty-four sheep, one goat and several cattle—twenty-four other sheep, one goat and several other cattle are going to be left without inoculation—then, at the appointed time, I am going to inject *all* of the beasts with the most deadly virulent culture of anthrax bacilli that we have. The vaccinated animals will be perfectly protected—the not-vaccinated ones will die in two days of course." Pasteur sounded as confident as an astronomer predicting an eclipse of the sun. . . .

"But, master, you know this work is so delicate—we *cannot* be absolutely sure of our vaccines—they may kill some of the sheep we try to protect—"

"WHAT WORKED WITH FOURTEEN SHEEP IN OUR LABORATORY WILL WORK WITH FIFTY AT MELUN!" Pasteur roared at them. For him just then, there was no such thing as a mysterious tricky nature, an unknown full of failures and surprises—the misty Infinite was as simple as two plus two makes four to him just then. So there was nothing for Roux and Chamberland to do but to roll up their sleeves and get the vaccines ready.

The day for the first injections came at last. Their bottles and syringes were ready, their flasks were carefully labeled—"Be sure not to mix up the first and second vaccine, boys!" shouted Pasteur, full of a gay confidence, as they left the Rue d'Ulm for the train. As they came on the field at Pouilly-le-Fort, and strode toward the sheds that held the forty-eight sheep, two goats and several cattle, Pasteur marched into the arena like a matador, and bowed severely to the crowd. There were senators of the Republic there, and scientists and horse doctors and dignitaries, and hundreds of farmers; and as Pasteur walked among them with his little limp—it was however a sort of jaunty limp—they cheered him mightily, many of them, and some of them snickered.

And there was a flock of newspaper men there, including the now almost legendary de Blowitz, of the London *Times*.

The sheep, fine healthy beasts, were herded into a clear space; Roux and Chamberland lighted their alcohol lamps and gingerly unpacked their glass syringes and shot five drops of the first vaccine—the anthrax bacilli that would kill mice but leave guinea-pigs alive, into the thighs of twenty-four of the sheep, one of the goats, and half of the cattle. The beasts got up and shook themselves and were labeled by a little gouge punched out of their ears. Then the audience repaired to a shed where Pasteur harangued them for half an hour—telling them simply but with a kind of dramatic portentousness of these new vaccinations and the hopes they held out for suffering men.

Twelve days went by and the show was repeated. The crowd was there once more and the second vaccine—the stronger one whose bacilli had the power of killing guinea-pigs but not rabbits—was injected, and the animals bore up beautifully under it and scampered about as healthy sheep, goats and cattle should do. The time for the fatal final test drew near; the very air of the little laboratory became finicky; the taut workers snapped at each other across the Bunsen flames. Pasteur was never so appallingly quiet—and the bottle washers fairly jumped across the room to fill his growled orders. Every day Thuillier, Pasteur's new youngest assistant, went out to the farm to put his thermometer carefully under the tails of the inoculated animals to see if they had fever— but thank God, every one of them was standing up beautifully under the heavy dose of the vaccine that was not quite murderous enough to kill rabbits.

While the heads of Roux and Chamberland turned several hairs grayer, Pasteur kept his confidence, and he wrote, with his old charmingly candid opinion of himself: "If success is complete, this will be one of the finest examples of applied science in this country, consecrating one of the greatest and most fruitful discoveries."

His friends shook their heads and lifted their shoulders and murmured: "Napoleonic, my dear Pasteur," and Pasteur did not deny it.

4

Then on the fateful thirty-first of May all of the forty-eight sheep, two goats, and several cattle—those that were vaccinated and those to which nothing whatever had been done—all of these received a surely fatal dose of virulent anthrax bugs. Roux got down on his knees in the dirt, surrounded by his alcohol lamps and bottles of deadly virus, and awed the crowd by his cool flawless shooting of the poisonous stuff into the more than sixty animals.

With his whole scientific reputation trusted to this one delicate test, realizing at last that he had done the brave but terribly rash thing of letting a frivolous public judge his science Pasteur rolled and tossed around in his bed and got up fifty times that night. He said absolutely nothing when Madame Pasteur tried to encourage him and told him, "Now now everything will come out all right"; he sulked in and out of the laboratory; there is no record of it, but without a doubt he prayed. . . .

Pasteur did not fancy going up in balloons and he would not fight duels—but no one can question his absolute gameness when he let the horse doctors get him into this dangerous test.

The crowd that came to judge Pasteur on the famous second day of June, 1881, made the previous ones look like mere assemblages at country baseball games. General Councilors were here to-day as well as senators; magnificoes turned out to see this show—tremendous dignitaries who only exhibited themselves to the public at the weddings and funerals of kings and princes. And the newspaper reporters clustered around the famous de Blowitz.

At two o'clock Pasteur and his cohorts marched upon the field and this time there were no snickers, but only a mighty bellowing of hurrahs. Not one of the twenty-four vaccinated sheep—though two days before millions of deadly germs had taken residence under their hides—not one of these sheep, I say, had so much as a trace of fever. They ate and frisked about as if they had never been within a thousand miles of an anthrax bacillus.

But the unprotected, the not vaccinated beasts—alas—there they lay in a tragic row, twenty-two out of twenty-four of them; and the remaining two were staggering about, at grips with that last inexorable, always victorious enemy of all living things. Ominous black blood oozed from their mouths and noses.

"See! There goes another one of those sheep that Pasteur did not vaccinate!" shouted an awed horse doctor.

5

The Bible does not go into details about what the great wedding crowd thought of Jesus when he turned water into wine, but Pasteur, that second of June, was the impresario of a modern miracle as amazing as any of the marvels wrought by the Man of Galilee, and that day Pasteur's whole audience—who many of them had been snickering skeptics—bowed down before this excitable little half-paralyzed man who could so perfectly protect living creatures from the deadly stings of sub-visible invaders. To me this beautiful experiment at Pouilly-le-Fort is an utterly strange event in the history of man's fight against relentless nature. There is no record of Prometheus bringing the precious fire to mankind amid applause; Galileo was actually clapped in prison for those searchings that have done more than any other to transform the world. We do not even know the names of those completely anonymous geniuses who first built the wheel and invented sails and thought to tame a horse.

6

But here stood Louis Pasteur, while his twenty-four immune sheep scampered about among the carcasses of the same number of pitiful dead ones, here stood this man, I say, in a gruesomely gorgeous stage-setting of an immortal drama, and all the world was there to see and to record and to be converted to his own faith in his passionate fight against needless death.

Now the experiment turned into the likeness of a revival. Doctor Biot, a healer in horses who had been one of the most sarcastic of the Pasteur-baiters, rushed up to him as the last of the not-vaccinated sheep was dying, and cried: "Inoculate me with your vaccines, Mr. Pasteur—just as you have done to those sheep you have saved so wonderfully—Then I will submit to the injection of the murderous virus! All men must be convinced of this marvelous discovery!"

"It is true," said another humbled enemy, "that I have made jokes about microbes, but I am a repentant sinner!"

"Well, allow me to remind you of the words of the Gospel," Pasteur answered him. "Joy shall be in heaven over one sinner that repenteth, more than over ninety and nine just persons that need no repentance."

The great de Blowitz cheered and rushed off to file his telegram to the London *Times* and to the newspapers of the world: "The experiment at Pouilly-le-Fort is a perfect, an unprecedented success."

The world received this news and waited, confusedly believing that Pasteur was a kind of Messiah who was going to lift from men the burden of all suffering. France went wild and called him her greatest son and conferred on him the Grand Cordon of the Legion of Honor. Agricultural societies, horse doctors, poor farmers whose fields were cursed with the poisonous virus of anthrax—all these sent telegrams begging him for thousands of doses of the life-saving vaccine. And

Pasteur, with Roux and Chamberland and Thuillier, responded to them with a magnificent disregard of their own health—and of science. For Pasteur, poet that he was, had more faith than the wildest of his new converts in this experiment.

In answer to these telegrams Pasteur turned the little laboratory in the Rue d'Ulm into a vaccine factory—huge kettles bubbled and simmered with the broth in which the tame, the life-saving, anthrax bacilli were to grow. Delicately—but so frantically that it was not quite delicate enough—Roux and Chamberland worked at weakening the murderous bacilli just enough to make the sheep of France a little sick, but not too sick from anthrax. Then all of them sweat at pouring numerous gallons of this bacillus-swarming soup which was the vaccine, into little bottles, a few ounces to each bottle, into clean bottles that had to be absolutely free from all other germs. And they had to do this subtle job without any proper apparatus whatever. I marvel that Pasteur ever attempted it; surely there never has been such blind confidence raised by one clear—but Lord! it might be simply a lucky—experiment.

In moments snatched from this making of vaccine Roux and Chamberland and Thuillier scurried up and down the land of France, and even to Hungary. They inoculated two hundred sheep in this place and five hundred and seventy-six in that—in less than a year hundreds of thousands of beasts had got this life-saving stuff. These wandering vaccinators would drag themselves back into the laboratory from their hard trips, they would get back to Paris probably wanting to get a few drinks or spend an evening with a pretty girl or loaf over a pipe—but Pasteur could not stand the smell of tobacco smoke, and as for wine and women, were not the sheep of France literally baa-ing to be saved? So these young men who were slaves of this battler whose one insane thought was "find-the-microbe-kill-the-microbe"—these faithful fellows took off their coats and peered at anthrax bacilli through the microscopes until their eye rims got red and their eyelashes fell out. In the middle of this work—with the farmers of France yelling for more vaccine—they began to have strange troubles: contaminating germs that had no business there began to pop up among the anthrax bacilli; all at once a weak vaccine that should have just killed a mouse began to knock off large rabbits. . . . Then, just as the scientific desperadoes got these messes straightened out, Pasteur would come in, nagging at them, fuming, fussing because they took so long at their experiments.

He wanted to try to find the deadly virus of hydrophobia.

And now at night the chittering of the guinea-pigs and the scurrying fights of the buck-rabbits in their cages were drowned by the eerie noise of mad dogs howling—sinister howls that kept Roux and Chamberland and Thuillier from sleep. . . . What would Pasteur ever have done—he surely would never have got far in his fight with the messengers of death—without those fellows Roux and Chamberland and Thuillier?

Gradually, it was hardly a year after the miracle of Pouilly-le-Fort, it began to be evident that Pasteur, though a most original microbe hunter, was not an infallible God. Disturbing letters began to pile up on his desk; complaints from Montpothier and a dozen towns of France, and from Packisch and Kapuvar in Hungary. Sheep were dying from anthrax—not natural anthrax they had picked up in dangerous fields, but anthrax they had got from those vaccines that were meant to save them! From other places came sinister stories of how the vaccine had failed to work—the vaccine had been paid for, whole flocks of sheep had been injected, the farmers had gone to bed breathing Thank-God-For-Our-Great-Man-Pasteur, only to wake up in the morning to find their fields littered with the carcasses of dead sheep, and these sheep—which ought to have been immune—had died from the lurking anthrax spores that lay in their fields. . . .

Pasteur began to hate to open his letters; he wanted to stop his ears against snickers that sounded from around corners, and then—the worst thing that could possibly happen—came a cold terribly exact scientific report from the laboratory of that nasty little German Koch in Berlin, and this report ripped the practicalness of the anthrax vaccine to tatters. Pasteur knew that Koch was the most accurate microbe hunter in the world.

There is no doubt that Pasteur lost some sleep from this aftermath of his glorious discovery, but, God rest him, he was a gallant man. It was not in him to admit, either to the public or to himself, that his sweeping claims were wrong.

"Have not *I* said that my vaccines made sheep a little sick with anthrax, but never killed them, and protected them perfectly? Well, I must stick to that," you can hear him mutter between his teeth.

What a searcher this Pasteur was, and yet how little of that fine selfless candor of Socrates or Rabelais is to be found in him. But he is not in any way to be blamed for that, for those two last were only, in their way, looking for truth, while Pasteur's work carried him more and more into the frantic business of saving lives, and in this matter truth is not of the first importance. . . .

In 1882, while his desk was loaded with reports of disasters, Pasteur went to Geneva, and there before the cream of disease-fighters of the world he gave a thrilling speech, subject: "How to guard living creatures from virulent maladies by injecting them with weakened microbes." Pasteur assured them that: "The general principles have been found and one cannot refuse to believe that the future is rich with the greatest hopes."

"We are all animated with a superior passion, the passion for progress and for truth!" he shouted—but unhappily he said no word about those numerous occasions when his vaccine had killed sheep instead of protecting them.

At this meeting Robert Koch sat blinking at Pasteur behind his gold-rimmed spectacles and smiling under his weedy beard at such an unscientific inspirational address. Pasteur seemed to feel something hanging over him, and he challenged Koch to argue with him publicly—knowing that Koch was a much better microbe hunter than an argufier. "I will content myself with replying to Mr. Pasteur's address in a written paper, in the near future," said Koch—who coughed, and sat down.

In a little while this reply appeared. It was dreadful. In this serio-comic answer Dr. Koch began by remarking that he had obtained some of this precious so-called anthrax vaccine from the agent of Mr. Pasteur.

Did Mr. Pasteur say that his *first* vaccine would kill mice, but not guinea-pigs? Dr. Koch had tested it, and it wouldn't even kill mice. But some queer samples of it killed sheep!

Did Mr. Pasteur maintain that his *second* vaccine killed guinea-pigs but not rabbits? Dr. Koch had carefully tested this one too, and found that it often killed rabbits very promptly —and sometimes sheep, poor beasts! which Mr. Pasteur claimed it would guard from death.

Did Mr. Pasteur really believe that his vaccines were really pure cultivations containing nothing but anthrax microbes? Dr. Koch had studied them carefully and found them to be veritable menageries of hideous scum-forming bacilli and strange cocci and other foreign creatures that had no business there.

Finally, was Mr. Pasteur really burning so with a passion for truth? Then why hadn't he told of the bad results as well as the good ones, that had followed the wholesale use of his vaccine?

"Such goings-on are perhaps suitable for the advertising of a business house, but science should reject them vigorously," finished Koch, drily, devastatingly.

Then Pasteur went through the roof and answered Koch's cool facts in an amazing paper with arguments that would not have fooled the jury of a country debating society. Did Koch dare to make believe that Pasteur's vaccines were full of con- taminating microbes? "For twenty years before Koch's scien- tific birth in 1876, it has been my one occupation to isolate and grow microbes in a pure state, and therefore Koch's in- sinuation that I do not know how to make pure cultivations cannot be taken seriously!" shouted Pasteur.

The French nation, even the great men of the nation, pa- triotically refused to believe that Koch had demoted their hero from the rank of God of Science—what could you expect from a German anyway?—and they promptly elected Pasteur to the *Académie Française*, the ultimate honor to bestow on a Frenchman. And on the day of Pasteur's admission this fiery yes-man was welcomed to his place among the Immortal Forty by the skeptical genius, Ernest Renan, the author who had changed Jesus from a God into a good human being, a man who could forgive everything because he understood everything. Renan knew that even if Pasteur sometimes did suppress the truth, he was still sufficiently marvelous. Renan was not a scientist but he was wise enough to know that Pasteur had done a wonderful thing when he showed that weak bugs may protect living beings against virulent ones—even if they would not do it one hundred times out of one hundred.

Regard these two fantastically opposite men facing each other on this solemn day. Pasteur the go-getter, an energetic fighter full of a mixture of faiths that interfered, sometimes, with ultimate—and maybe ugly—truth. And talking to him loftily sits the untroubled Renan with the massiveness of Mount Everest, such a dreadful skeptic that he probably was never quite convinced that he was himself alive, so firmly doubting the value of doing anything that he had become one of the fattest men in France.

Renan called Pasteur a genius and compared him to some of the greatest men that ever lived and then gave the excited, paralyzed, gray-haired microbe hunter this mild admonition:

"Truth, Sir, is a great coquette; she will not be sought with too much passion, but often is most amenable to indifference. She escapes when apparently caught, but gives herself up if patiently waited for; revealing herself after farewells have been said, but inexorable when loved with too much fervor."

Surely Renan was too wise to think that his lovely words would ever change Pasteur one jot from the headlong untruthful hunter after truth that he was. But just the same, these words sum up the fundamental sadness of Pasteur's life, they tell of the crown of thorns that madmen wear whose dream it is to change a world in the little seventy years they are allowed to live.

7

And now Pasteur began—God knows why—to stick little hollow glass tubes into the gaping mouths of dogs writhing mad with rabies. While two servants pried apart and held open the jowls of a powerful bulldog, Pasteur stuck his beard within a couple of inches of those fangs whose snap meant the worst of deaths, and, sprinkled sometimes with a maybe fatal spray, he sucked up the froth into his tube—to get a specimen in which to hunt for the microbe of hydrophobia. I wish to forget, now, everything that I have said about his showmanship, his unsearcherlike go-gettings. This business of his gray eyes looking that bulldog in the mouth—this was no grandstand stuff.

Why did Pasteur set out to trap the germ of rabies? That is a mystery, because there were a dozen other serious diseases, just then, whose microbes had not yet been found, diseases that killed many more people than rabies had ever put to death, diseases that were not nearly so surely deadly to an adventurous experimenter as rabies would be—if one of those dogs should get loose. . . .

It must have been the artist, the poet in him that urged him on to this most hard and dangerous hunting, for Pasteur himself said: "I have always been haunted by the cries of those victims of the mad wolf that came down the street of Arbois when I was a little boy. . . ." Pasteur knew the way the yells of a mad dog curdle the blood of every one. He remembered that less than a hundred years before in France, laws had to be passed against the poisoning, the strangling, the shooting of wretched people whom frightened fellow-townsmen just suspected of having rabies. Doubtless he saw himself the deliverer of men from such crazy fear—such hopeless suffering.

And then, in this most magnificent and truest of all his searchings, Pasteur started out, as he so often did, by making mistakes. In the saliva of a little child dying from hydrophobia he discovered a strange motionless germ that he gave the unscientific name of "microbe-like-an-eight." He read papers at the Academy that hinted about this figure-eight germ having something to do with the mysterious cause of hydrophobia. But in a little while this trail proved to be a blind one, for with Roux and Chamberland he found—after he had settled down and got his teeth into this search—that this eight-microbe could be found in the mouths of many healthy people who had never been anywhere near a mad dog.

Presently, late in 1882, he ran on to his first clew. "Mad dogs are scarce just now, old Bourrel the veterinarian brings me very few of them, and people with hydrophobia are still harder to get hold of—we've got to produce this rabies in animals in our laboratory and keep it going there—otherwise we won't be able to go on studying it steadily," he pondered.

He was more than sixty, and he was tired.

Then one day, a lassoed mad dog was brought into the laboratory; dangerously he was slid into a big cage with healthy dogs and allowed to bite them. Roux and Chamberland fished froth out of the mouth of this mad beast and sucked it up into syringes and injected this stuff into rabbits and guinea-pigs. Then they waited eagerly to see this menagerie develop the first signs of madness. Sometimes—alas—the experiment worked, but other very irritating times it did not; four healthy dogs had been bitten and six weeks later they came in one morning to find two of these creatures lashing about their cages, howling—but for months after that the other two showed no sign of rabies; there was no rime or reason to this business, no regularity, confound it! this was not *science!* And it was the same with the guinea-pigs and rabbits: two of the rabbits might drag out their hind legs with a paralysis—then die in dreadful convulsions, but the other four would go on chewing their greens as if there were no mad-dog virus within a million miles of them.

Then one day a little idea came to Pasteur, and he hurried to tell it to Roux.

"This rabies virus that gets into people by bites, it settles in their brains and spinal cords. . . . All the symptoms of hydrophobia show that it's the nervous system that this virus—this bug we can't find—attacks. . . .

"That's where we have to look for the unknown microbe . . . that's where we can grow it maybe, even without seeing it . . . maybe we could use the living animal's brain instead of a bottle of soup . . . a funny culture-bottle that would be, but. . . .

"When we inject it under the skin—the virus may get lost in the body before it can travel to the brain—if I could only stick it right into a dog's brain . . . !"

Roux listened to these dreamings of Pasteur, he listened bright-eyed to these fantastic imaginings. . . . Another man than Roux might have thought Pasteur completely crazy. . . . The brain of a dog or rabbit instead of a bottle of broth, indeed! What nonsense! But not to Roux!

"But why not put the virus right into a dog's brain, master, I can trephine a dog—I can drill a little hole in his skull—without hurting him—without damaging his brain at all . . . it would be easy . . ." said Roux.

Pasteur shut Roux up, furiously. He was no doctor, and he did not know that surgeons can do this operation on human beings even, quite safely. "What! bore a hole right through a dog's skull—why, you'd hurt the poor beast terribly . . . you would damage his brain . . . you would paralyze him . . . No! I will not permit it!"

So near was Pasteur, by reason of his tender-heartedness, so close was he to failing completely in winning to the most marvelous of his gifts to men. He quailed before the stern experiment that his weird idea demanded. But Roux—the faithful, the now almost forgotten Roux—saved him by disobeying him.

For, a few days later when Pasteur left the laboratory to go to some meeting or other, Roux took a healthy dog, put him easily out of pain with a little chloroform, and bored a hole in the beast's head and exposed his palpitating, living brain. Then up into a syringe he drew a little bit of the ground-up brain of a dog just dead with rabies: "This stuff must be swarming with those rabies microbes that are maybe too small for us to see," he pondered; and through the hole in the sleeping dog's skull went the needle of the syringe, and into the living brain Roux slowly, gently shot the deadly rabid stuff. . . .

Next morning Roux told Pasteur about it—"What!" shouted Pasteur. "Where is the poor creature . . . he must be dying . . . paralyzed. . . ."

But Roux was already down the stairs, and in an instant he was back, his operated dog prancing in ahead of him, jumping gayly against Pasteur, sniffing 'round among the old broth bottles under the laboratory benches. Then Pasteur realized Roux's cleverness—and the new road of experiment that lay before him, and though he was not fond of dogs, his joy made him fuss over this one: "Good dog, excellent beast!" Pasteur said, and dreamed: "This beast will show that my idea will work. . . ."

Sure enough, less than two weeks later the good creature began to howl mournful cries and tear up his bed and gnaw at his cage—and in a few days more he was dead, and this brute died, as you will see, so that thousands of mankind might live.

Now Pasteur and Roux and Chamberland had a sure way, that worked one hundred times out of one hundred, of giving rabies to their dogs and guinea-pigs and rabbits. "We cannot find the microbe—surely it must be too tiny for the strongest microscope to show us—there's no way to grow it in flasks of soup . . . but we can keep it alive—this deadly virus—in the brains of rabbits . . . that is the only way to grow it," you can hear Pasteur telling Roux and Chamberland.

Never was there a more fantastic experiment in all of microbe hunting, or in any science, for that matter; never was there a more unscientific feat of science than this struggling, by Pasteur and his boys, with a microbe they couldn't see—a weird bug of whose existence they only knew by its invisible growth in the living brains and spinal cords of an endless succession of rabbits and guinea-pigs and dogs. Their only knowledge that there was such a thing as the microbe of rabies was the convulsive death of the rabbits they injected, and the fearful cries of their trephined dogs. . . .

Then Pasteur and his assistants started on their outlandish —any wise man would say their impossible—adventure of taming this vicious virus that they could not see. There were little interruptions; Roux went with Thuillier to fight the cholera in Egypt and there, you will remember, Thuillier died; and Pasteur went out into the rural pig-sties of France to discover the microbe and find a vaccine against a disease that was just then murdering French swine. But Pasteur stopped getting entangled in those vulgar arguments which were so often to his discredit, and the three of them locked themselves in their laboratory in the Rue d'Ulm with their poor paralyzed and dangerous animals. They sweat through endless experiments.

Pasteur mounted guard over his young men and kept their backs bent over their benches as if they were some higher kind of galley slave. He watched their perilous experiments with one eye and kept the other on the glass door of the workroom, and when he saw some of Roux's and Chamberland's friends approaching, to ask them maybe to come out for a glass of beer on the terrace of a near-by café, the master would hurry out and tell the interlopers: "No. No! Not now! Cannot you see? They are busy—it is a most important experiment they are doing!"

Months—gray months went by during which it seemed to all of them that there was no possible way of weakening the invisible virus of rabies. . . . One hundred animals, alas, out of every hundred that they injected—died. You would think that Roux and Chamberland, still youngsters, would have been the indomitable ones, the never-say-die men of this desperate crew. But on the contrary!

"It's no go, master," said they, making limp waves of their hands toward the cages with their paralyzed beasts—toward the tangled jungles of useless tubes and bottles. . . .

Then Pasteur's eyebrows cocked at them, and his thinning

gray hair seemed to stiffen: "Do the same experiment over again—no matter if it failed last time—it may look foolish to you, but the important thing is not to leave the subject!" Pasteur shouted, in a fury. So it was that this man scolded his monkish disciples and prodded them to do useless tests over and over and over—with no reasons, with complete lack of logic. With every fact against him Pasteur searched and tried and failed and tried again with that insane neglect of common sense that sometimes turns hopeless causes into victories.

Indeed, why wasn't this setting out to tame the hydrophobia virus—why wasn't it a nonsensical wild-goose chase? There was in all human history no single record of any man or beast getting better from this horrible malady, once the symptoms had declared themselves, once the mysterious messengers of evil had wormed their unseen way into the spinal cord and brain. It was this kind of murderous stuff that Pasteur and his men balanced on the tips of their knives, sucked up into their glass pipettes within an inch from the lips—stuff that was separated from their mouths by a thin little wisp of cotton. . . .

Then, one exciting day, the first sweet music of encouragement came to these gropers in the dark—one of their dogs inoculated with the surely fatal stuff from a rabid rabbit's brain—this dog came down with his weird barkings and portentous shiverings and slatherings—and then miraculously got completely better! Excitedly, a few weeks later, they shot this first of all recovered beasts with a deadly virus, directly into his brain they injected the wee murderers. The little wound on his head healed quickly—anxiously Pasteur waited for his doomful symptoms to come on him, but these signs never came. For months the dog romped about his cage. He was absolutely immune!

"Now we know it—we know we have a chance. . . . When a beast once has rabies and gets better from it, there will be no recurrence. . . . We must find a way to *tame* the virus now," said Pasteur to his men, who agreed, but were perfectly certain that there was no way to tame that virus.

But Pasteur began inventing experiments that no god would have attempted; his desk was strewn with hieroglyphic scrawls of them. And at eleven in the morning, when the records of the results of the day before had been carefully put down, he would call Roux and Chamberland, and to them he would read off some wild plan for groping after this unseen unreachable virus—some fantastic plan for getting his fingers on it *inside* the body of a rabbit—to weaken it.

"Try this experiment to-day!" Pasteur would tell them.

"But that is technically impossible!" they protested.

"No matter—plan it any way you wish, provided you do it well," Pasteur replied. (He was, those days, like old Ludwig van Beethoven writing unplayable horn parts for his symphonies—and then miraculously discovering hornblowers to play those parts.) For, one way or another, the ingenious Roux and Chamberland devised tricks to do those crazy experiments. . . .

And at last they found a way of weakening the savage hydrophobia virus—by taking out a little section of the spinal cord of a rabbit dead of rabies, and hanging this bit of deadly stuff up to dry in a germ-proof bottle for fourteen days. This shriveled bit of nervous tissue that had once been so deadly they shot into the brains of healthy dogs—and those dogs did not die. . . .

"The virus is dead—or better still very much weakened," said Pasteur, jumping at the latter conclusion with no sense or reason. "Now we'll try drying other pieces of virulent stuff for twelve days—ten days—eight days—six days, and see if we can't just give our dogs a *little* rabies . . . then they ought to be immune. . . ."

Savagely they fell to this long will o' the wisp of an experiment. For fourteen days Pasteur walked up and down the bottle and microscope and cage-strewn unearthly workshop and grumbled and fretted and made scrawls in that everlasting notebook of his. The first day the dogs were dosed with the weakened—the almost extinct virus that had been dried for fourteen days; the second day they received a shot of the slightly stronger nerve stuff that had been thirteen days in its bottle; and so on until the fourteenth day—when each beast was injected with one-day-dried virus that would have surely killed a not-inoculated animal.

For weeks they waited—hair graying again—for signs of rabies in these animals, but none ever came. They were happy, these ghoulish fighters of death! Their clumsy terrible fourteen vaccinations had not hurt the dogs—but were they immune?

Pasteur dreaded it—if this failed all of these years of work had gone for nothing, and "I am getting old, old . . ." you can hear him whispering to himself. But the test had to be made. Would the dogs stand an injection of the most deadly rabid virus—right into their brains—a business that killed an ordinary dog one hundred times out of one hundred?

Then one day Roux bored little holes through the skulls of two vaccinated dogs—and two not vaccinated ones: and into all four went a heavy dose of the most virulent virus. . . .

One month later, Pasteur and his men, at the end of three years of work, knew that victory over hydrophobia was in their hands. For, while the two vaccinated dogs romped and sniffed about their cages with never a sign of anything ailing them— the two that had not received the fourteen protective doses of dried rabbit's brain—these two had howled their last howls and died of rabies.

Now immediately—the life-saver in this man was always downing the mere searcher—Pasteur's head buzzed with plans to wipe hydrophobia from the earth, he had a hundred foolish projects, and he walked in a brown world of thought, in a mist of plans that Roux and Chamberland, and not even Madame Pasteur could penetrate. It was 1884, and when Pasteur forgot their wedding anniversary, the long-suffering lady wrote to her daughter:

"Your father is absorbed in his thoughts, talks little, sleeps little, rises at dawn, and, in one word, continues the life I began with him this day thirty-five years ago."

At first Pasteur thought of shooting his weakened rabies

virus into all the dogs of France in one stupendous Napoleonic series of injections: "We must remember that no human being is ever attacked with rabies except after being bitten by a rabid dog. . . . Now if we wipe it out of dogs with our vaccine . . ." he suggested to the famous veterinarian, Nocard, who laughed, and shook his head.

"There are more than a hundred thousand dogs and hounds and puppies in the city of Paris alone," Nocard told him, "and more than two million, five hundred thousand dogs in all of France—and if each of these brutes had to get fourteen shots of your vaccine fourteen days in a row . . . where would you get the men? Where would you get the time? Where the devil would you get the rabbits? Where would you get sick spinal cord enough to make one-thousandth enough vaccine?"

Then finally there dawned on Pasteur a simple way out of his trouble: "It's not the dogs we must give our fourteen doses of vaccine," he pondered, "it's the human beings that have been bitten by mad dogs. . . ."

"How easy! . . . After a person has been bitten by a mad dog, it is always weeks before the disease develops in him. . . . The virus has to crawl all the way from the bite to the brain. . . . While that is going on we can shoot in our fourteen doses . . . and protect him!" and hurriedly Pasteur called Roux and Chamberland together, to try it on the dogs first.

They put mad dogs in cages with healthy ones, and the mad dogs bit the normal ones.

Roux injected virulent stuff from rabid rabbits into the brains of other healthy dogs.

Then they gave these beasts, certain to die if they were left alone—they shot the fourteen stronger and stronger doses of vaccine into them. It was an unheard-of triumph! For every one of these creatures lived—threw off perfectly, mysteriously, the attacks of their unseen assassins, and Pasteur—who had had a bitter experience with his anthrax inoculations—asked that all of his experiments be checked by a commission of the best medical men of France, and at the end of these severe experiments the commission announced:

"Once a dog is made immune with the gradually more virulent spinal cords of rabbits dead of rabies, nothing on earth can give him the disease."

From all over the world came letters, urgent telegrams, from physicians, from poor fathers and mothers who were waiting terror-smitten for their children, mangled by mad dogs, to die—frantic messages poured in on Pasteur, begging him to send them his vaccine to use on threatened humans. Even the magnificent Emperor of Brazil condescended to write Pasteur, begging him . . .

And you may guess how Pasteur was worried! This was no affair like anthrax, where, if the vaccine was a little, just a shade too strong, a few sheep would die. Here a slip meant the lives of babies. . . . Never was any microbe hunter faced with a worse riddle. "Not a single one of all my dogs has ever died from the vaccine," Pasteur pondered. "All of the bitten ones have been perfectly protected by it. . . . It must work the same way on humans—it *must* . . . but . . ."

And then sleep once more was not to be had by this poor searcher who had made a too wonderful discovery. . . . Horrid pictures of babies crying for the water their strangled throats would not let them drink—children killed by his own hands —such visions floated before him in the dark. . . .

For a moment the actor, the maker of grand theatric gestures, rose in him again: "I am much inclined to begin on myself—inoculating myself with rabies, and then arresting the consequences; for I am beginning to feel very sure of my results," he wrote to his old friend, Jules Verçel.

At last, mercifully, the worried Mrs. Meister from Meissengott in Alsace took the dreadful decision out of Pasteur's unsure hands. This woman came crying into the laboratory, leading her nine-year-old boy, Joseph, gashed in fourteen places two days before by a mad dog. He was a pitifully whimpering, scared boy—hardly able to walk.

"Save my little boy—Mr. Pasteur," this woman begged him.

Pasteur told the woman to come back at five in the evening,

and meanwhile he went to see the two physicians, Vulpian and Grancher—admirers who had been in his laboratory, who had seen the perfect way in which Pasteur could guard dogs from rabies after they had been terribly bitten. That evening they went with him to see the boy, and when Vulpian saw the angry festering wounds he urged Pasteur to start his inoculations: "Go ahead," said Vulpian, "if you do nothing it is almost sure that he will die."

And that night of July 6, 1885, they made the first injection of the weakened microbes of hydrophobia into a human being. Then, day after day, the boy Meister went without a hitch through his fourteen injections—which were only slight pricks of the hypodermic needle into his skin.

And the boy went home to Alsace and had never a sign of that dreadful disease.

Then all fears left Pasteur—it was very much like the case of that first dog that Roux had injected years before, against the master's wishes. So it was now with human beings; once little Meister came through unhurt, Pasteur shouted to the world that he was prepared to guard the people of the world from hydrophobia. This one case had completely chased his fears, his doubts—those vivid but not very deep-lying doubts of the artist that was in Louis Pasteur.

The tortured bitten people of the world began to pour into the laboratory of the miracle-man of the Rue d'Ulm. Research for a moment came to an end in the messy small suite of rooms, while Pasteur and Roux and Chamberland sorted out polyglot crowds of mangled ones, babbling in a score of tongues: "Pasteur—save us!"

And this man who was no physician—who used to say with proud irony: "I am only a chemist,"—this man of science who all his life had wrangled bitterly with doctors, answered these cries and saved them. He shot his complicated, illogical fourteen doses of partly weakened germs of rabies—unknown microbes of rabies—into them and sent these people healthy back to the four corners of the earth.

From Smolensk in Russia came nineteen peasants, moujiks

who had been set upon by a mad wolf nineteen days before, and five of them were so terribly mangled they could not walk at all, and had to be taken to the Hotel Dieu. Strange figures in fur caps they came, saying: "Pasteur—Pasteur," and this was the only word of French they knew.

Then Paris went mad—as only Paris can—with excited concern about these bitten Russians who must surely die—it was so long since they had been attacked—and the town talked of nothing else while Pasteur and his men started their injections. The chances of getting hydrophobia from the bites of mad wolves are eight out of ten: out of these nineteen Russians, fifteen were sure to die. . . .

"Maybe," said every one, "they will all die—it is more than two weeks since they were attacked, poor fellows; the malady must have a terrible start, they have no chance. . . ." Such was the gabble of the Boulevards.

Perhaps, indeed, it was too late. Pasteur could not eat nor did he sleep at all. He took a terrible risk, and morning and night, twice as quickly as he had ever made the fourteen injections—twice a day to make up for lost time—he and his men shot the vaccine into the arms of the Russians.

And at last a great shout of pride went up for this man Pasteur, went up from the Parisians, and all of France and all the world raised a pæan of thanks to him—for the vaccine marvelously saved all but three of the doomed peasants. The moujiks returned to Russia and were welcomed with the kind of awe that greets the return of hopeless sick ones who have been healed at some miraculous shrine. And the Tsar of All the Russias sent Pasteur the diamond cross of Ste. Anne, and a hundred thousand francs to start the building of that house of microbe hunters in the Rue Dutot in Paris—that laboratory now called the Institut Pasteur. From all over the world—it was the kind of burst of generosity that only great disasters usually call out—from every country in the earth came money, piling up into millions of francs for the building of a laboratory in which Pasteur might have everything needed to track down other deadly microbes, to invent weapons against them. . . .

The laboratory was built, but Pasteur's own work was done; his triumph was too much for him; it was a kind of trigger, perhaps, that snapped the strain of forty years of never before heard-of ceaseless searching. He died in 1895 in a little house near the kennels where they now kept his rabid dogs, at Villeneuve l'Etang, just outside of Paris. His end was that of the devout Catholic, the mystic he had always been. In one hand he held a crucifix and in the other lay the hand of the most patient, obscure and important of his collaborators—Madame Pasteur. Around him, too, were Roux and Chamberland and those other searchers he had worn to tatters with his restless energy, those faithful ones he had abused, whom he had above all inspired; and these men who had risked their lives in the carrying out of his wild forays against death would now have died to save him, if they could.

That was the perfect end of this so human, so passionately imperfect hunter of microbes and saver of lives.

But there is another end of his career that I like to think of more—and that was the day, in 1892, of Pasteur's seventieth birthday—when a medal was given to him at a great meeting held to honor him, at the Sorbonne in Paris. Lister was there, and many other famous men from other nations, and in tier upon tier, above these magnificoes who sat in the seats of honor, were the young men of France—the students of the Sorbonne and the colleges and the high schools. There was a great buzz of young voices—all at once a hush, as Pasteur limped up the aisle, leaning on the arm of the President of the French Republic. And then—it is the kind of business that is usually pulled off to welcome generals and that kind of hero who has directed the futile butchering of thousands of enemies—the band of the Republican Guard blared out into a triumphal march.

Lister, the prince of surgeons, rose from his seat and hugged Pasteur and the gray-bearded important men and the boys in the top galleries cried and shook the walls with the roar of their cheering. At last the old microbe hunter gave his speech—the voice of the fierce arguments was gone and his son had to speak it for him—and his last words were a hymn of hope, not so much for the saving of life as a kind of religious cry for a new way of life for men. It was to the students, to the boys of the high schools he was calling:

". . . Do not let yourselves be tainted by a deprecating and barren skepticism, do not let yourselves be discouraged by the sadness of certain hours which pass over nations. Live in the serene peace of laboratories and libraries. Say to yourselves first: What have I done for my instruction? and, as you gradually advance, What have I done for my country? until the time comes when you may have the immense happiness of thinking that you have contributed in some way to the progress and good of humanity. . . ."

6

Roux and Behring

Massacre the Guinea-Pigs

1

It was to save babies that they killed so many guinea-pigs!

Emile Roux, the fanatical helper of Pasteur, in 1888 took up the tools his master had laid down, and started on searches of his own. In a little while he discovered a strange poison seeping from the bacillus of diphtheria—one ounce of the pure essence of this stuff was enough to kill seventy-five thousand big dogs. A few years later, while Robert Koch was bending under the abuse and curses of sad ones who had been disappointed by his supposed cure for consumption, Emil Behring, the poetical pupil of Koch, spied out a strange virtue, an unknown something in the blood of guinea-pigs. It could make that powerful diphtheria poison completely harmless. . . . These two Emils revived men's hopes after Koch's disaster, and once more people believed for a time that microbes were going to be turned from assassins into harmless little pets.

What experiments these two young men made to discover this diphtheria antitoxin! They went at it frantic to save lives; they groped at it among bizarre butcherings of countless guinea-pigs; in the evenings their laboratories were shambles like the battlefields of old days when soldiers were mangled by spears and pierced by arrows. Roux dug ghoulishly into the spleens of dead children—Behring bumped his nose in the darkness of his ignorance against facts the gods themselves could not have predicted. For each brilliant experiment these two had to pay with a thousand failures.

But they discovered the diphtheria antitoxin.

They never could have done it without the modest discovery of Frederick Loeffler. He was that microbe hunter whose mustache was so militaristic that he had to keep pulling it down to see through his microscope; he sat working at Koch's right hand in that brave time when the little master was tracking down the tubercle bacillus. It was in the early eighteen eighties, and diphtheria, which several times each hundred years seems to have violent ups and downs of viciousness— diphtheria was particularly murderous then. The wards of the hospitals for sick children were melancholy with a forlorn wailing; there were gurgling coughs foretelling suffocation; on the sad rows of narrow beds were white pillows framing small faces blue with the strangling grip of an unknown hand. Through these rooms walked doctors trying to conceal their hopelessness with cheerfulness; powerless they went from cot to cot—trying now and again to give a choking child its breath by pushing a tube into its membrane-plugged windpipe. . . .

Five out of ten of these cots sent their tenants to the morgue.

Below in the dead house toiled Frederick Loeffler, boiling knives, heating platinum wires red hot and with them lifting grayish stuff from the still throats of those bodies the doctors had failed to keep alive; and this stuff he put into slim tubes capped with white fluffs of cotton, or he painted it with dyes, which showed him, through his microscope, that there were queer bacilli shaped like Indian clubs in those throats, microbes which the dye painted with pretty blue dots and stripes and bars. In nearly every throat he discovered these strange bacilli; he hurried to show them to his master, Koch.

There is little doubt Koch led Loeffler by the hand in this discovery. "There is no use to jump at conclusions," you can hear Koch telling him. "You must grow these microbes pure —then you must inject the cultivations into animals. . . . If those beasts come down with a disease exactly like human diphtheria, then . . ." How could Loeffler have gone wrong, with that terribly pedantic, but careful, truth-hunting little czar of microbe hunters squinting at him from behind those eternal spectacles?

One dead child after another Loeffler examined; he poked into every part of each pitiful body; he stained a hundred different slices of every organ; he tried—and quickly succeeded —in growing those queer barred bacilli pure. But everywhere he searched, in every part of each body, he found no microbes—except in the membrane-cluttered throat. And always here, in every child but one or two, he came on those Indian club-shaped rods. "How can these few microbes, growing nowhere in the body but the throat—how can these few germs, staying in that one place, kill a child so quickly?" pondered Loeffler. "But I must follow Herr Koch's directions!" and he proceeded to shoot the germs of his pure cultivations into the windpipes of rabbits and beneath the skins of guinea-pigs. Quickly these animals died—in two or three days, like a child, or even more quickly—but the microbes, which Loeffler had shot into them in millions, could only be found at the spot where he had injected them. . . . And sometimes there were none to be found even here, or at best a few feeble ones hardly strong enough, you would think, to hurt a flea. . . .

"But how is it these few bacilli—sticking in one little corner of the body—how can they topple over a beast a million times larger than they are themselves?" asked Loeffler.

Never was there a more conscientious searcher than this Loeffler, nor one with less of a wild imagination to liven—or to spoil—his almost automatic exactness. He sat himself down; he wrote a careful scientific paper; it was modest, it was cold, it was not hopeful, it was a most unlawyer-like report reciting all of the fors and againsts on the question of whether or no this new bacillus was the cause of diphtheria. He leaned over backward to be honest—he put last the facts that were against it! "This microbe *may* be the cause," you can hear him mumbling as he wrote, "but in a few children dead of diphtheria I could not find these germs . . . none of my inoculated animals get paralysis as children do . . . what is most against me is that I've discovered this same microbe—it was vicious against guinea-pigs and rabbits too!—in the throat of a child with never a sign of diphtheria."

He even went so far as to underestimate the importance of his exact fine searching, but at the end of his treatise he gave a clew to the more imaginative Roux and Behring who came after him. A strange man, this Loeffler! Without seeming to be able to make a move to do it himself, he predicted what others must find:

"This bacillus stays on a little patch of dead tissue in the throat of a baby; it lurks on a little point under a guinea-pig's skin; it never swarms in millions—yet it kills! How?

"It must make a poison—a toxin that leaks out of it, sneaking from it to some vital spot in the body. Such a toxin must be found, in the organs of a dead child, in the carcass of a guinea-pig dead of the disease—yes—and in the broth where the bacillus grows so well. . . . The man finding this poison will prove what I have failed to demonstrate." Such was the dream Loeffler put into Roux's head. . . .

2

Four years later Loeffler's words came true—by what seemed an utterly silly, but what was surely a most fantastical experiment you would have thought could only result in drowning a guinea-pig. What a hectic microbe hunting went on in Paris just then! Pasteur, in a state of collapse after his triumph of the dog bite vaccine, was feebly superintending the building of his million-franc Institute in the Rue Dutot. The wild, half-charlatan Metchnikoff had come out of Odessa in Russia to belch quaint theories about how phagocytes gobble up malignant germs. Pasteurians were packing microscopes in satchels and hurrying to Saigon in Indo-China and to Australia to try to discover microbes of weird diseases that did not exist. Hopefully frantic women were burying Pasteur—he was too tired!—under letters begging him to save their children from a dozen horrid diseases.

"If you will," one woman wrote him, "you can surely find a remedy for the horrible disease called diphtheria. Our children, to whom we teach your name as a great benefactor, will owe their lives to you!"

Pasteur was absolutely done up, but Roux—and he was helped by the intrepid Yersin who afterward brilliantly discovered the germ of the black death—set out to try to find a way to wipe diphtheria from the earth. It wasn't a science—it was a crusade, this business. It was full of passion, of purpose; it lacked skillful lying-in-wait, and those long planned artistic ambushes you find in most discoveries. I will not say Emile Roux began his searching because of this pitiful note from that woman—but there is no doubt he worked to save rather than to know. From the old palsied master down to the most obscure bottle wiper, the men of this house in the Rue Dutot were humanitarians; they were saviors—and that is noble!— but this drove them sometimes into strange byways far off the road where you find truth. . . . And in spite of this Roux made a marvelous discovery.

Roux and Yersin went to the Hospital for Sick Children— diphtheria was playing hell with Paris—and here they ran on to the same bacillus Loeffler had found. They grew this microbe in flasks of broth, and did the regular accepted thing first, shooting great quantities of this soup into an assorted menagerie of unfortunate birds and quadrupeds who had to die without the satisfaction of knowing they were martyrs. It wasn't particularly enlightened searching, this, but almost from the tap of the gong, they stumbled on one of the proofs Loeffler had failed to find. Their diphtheria soup paralyzed rabbits! The stuff went into their veins; in a few days the delighted experimenters watched these beasts drag their hind legs limply after them; the palsy crept up their bodies to their front legs and shoulders—they died in a clammy, dreadful paralysis. . . .

"It hits rabbits just the way it does children," muttered Roux, full of a will to believe—"This bacillus must be the true cause of diphtheria. . . . I shall find the germ in these rabbits' bodies now!" And he clawed tissues out of a dozen corners of their carcasses; he made cultivations of their spleens and hearts—but never a bacillus! Only a few days before he had pumped a billion or so into them, each of them. Here they were, drawn and quartered, carved up and searched from their pink noses to the white under-side of their tails. And not a bacillus. What had killed them then?

Then Loeffler's prediction flashed over Roux: "It must be the germs make a poison, in this broth, to paralyze and kill these beasts . . ." he pondered.

For a while the searcher came uppermost in him. He forgot about possible savings of babies; he concentrated on vast butcheries of guinea-pigs and rabbits—he must prove that the diphtheria germ drips a toxin out of its wee body. . . . Together with Yersin he began a good unscientific fumbling at experiments; they were in the dark; there were no precedents nor any kind of knowledge to go by. No microbe hunter before them had ever separated a deadly poison (though Pasteur had once made something of a try at it) from the bodies of microbes. They were alone in the dark, Roux and Yersin—but they lighted matches. . . . "The bacilli *must* pour out a poison into the broth we grow them in—just as they pour it from their membrane in a child's throat into his blood!" Of course that last was not proved.

Then Roux stopped arguing in a circle. He searched. He worked with his hands. It was worse, this fumbling of his, than trying to get a stalled motor to go when you know nothing about internal combustion machinery. He took big glass bottles and put pure microbeless soup into them, and sowed pure cultivations of the diphtheria bacillus in this broth; into the incubating oven went the large-bellied bottles—"Now we will try separating the germs from the soup in which they grow," said Roux, after the bottle had ripened for four days. They rigged up a strange apparatus—it was a filter, shaped like a candle, only it was hollow, and made of fine porcelain that would let the soup through, but so tight-meshed that it would hold the tiniest bacilli back. With tongue-protruding care to keep themselves from being splashed with this deadly stuff, they poured the microbe-teeming broth around the candles held rigid in shiny glass cylinders. They fussed—maybe, or at least I hope so, with the blessed relief of profanity—but the broth wouldn't run through the porcelain. But at last they pushed it through with high air pressure—and finally they breathed easy, arranging little flasks full of a clear, amber-colored filtered fluid (it had never a germ in it) on their laboratory bench.

"This stuff should have the poison in it . . . the filter has held back all the microbes—but this stuff should kill our animals," muttered Roux. The laboratory buzzed with eager animal-boys getting ready the rabbits and guinea-pigs. Into the bellies of these beasts went the golden juice propelled from the syringe by Roux's deft hands. . . .

He became a murderer in his heart, this Emile Roux, and in his head as he came down to the laboratory each morning were half-mad wishes for the death of his beasts. "The stuff should be hitting them by now," you can hear him growling to Yersin, but they looked in vain for the ruffled hair, the dragging hind legs, the cold shivering bodies to tell them their wish was coming true.

It was beastly! All of this fussing with the delicate filter experiments—and the animals munched at the greens in their cages, they hopped about, males sniffed at females and engaged in those absurd scufflings with other males which guinea-pigs and rabbits hold to be necessary to the propagation of their kind. . . . Let these giants (who fed them well) inject more of this stuff into their veins, their bellies—poison? Imagination! It made them feel happy. . . .

Roux tried again. He shot bigger doses of his filtered soup

into the animals, other animals, still more animals. It was no go, there was no poison.

That is, for a merely sensible man there would have been no poison in the filtered soup that had stood in the incubator for four days. Hadn't enough animals been wasted trying it? But Roux (let all mothers and children and the gods caring for insane searchers bless him!) was no reasonable man just then. For a moment he had caught Pasteur's madness, his strange trick of knowing what all men thought wrong to be right, his flair for good impossible experiments. "There is a poison there!" you can hear that hawk-faced consumptive Roux shout to himself, to the dusty, bottle-loaded shelves of his laboratory, to the guinea-pigs who would have snickered—if they could have—at his earnest futile efforts to murder them. "There must be a poison in this soup where the diphtheria germs have grown—else why should those rabbits have died?"

Then—I have told scientific searchers about this and they have held their noses at such an experiment—Roux nearly drowned a guinea-pig. For weeks he had been injecting more and more of his filtered soup, but now (it was like facing a night on a park bench with your last dime on the two dice) he injected thirty times as much! Not even Pasteur would have risked such an outlandish dose—thirty-five cubic centimeters Roux shot under the guinea-pig's skin and you would expect that much water would kill such a little beast. If he died it would mean nothing. . . . But into the belly of a guinea-pig and into the ear-vein of a rabbit went this ocean of filtered juice—it was as if he had put a bucketful of it into the veins of a middle-sized man.

But that was the way Roux carved his name on those tablets which men while they are on earth must never allow to crumble; for, though the rabbit and the guinea-pig stood the mere bulk of the microbe-less broth very well, and appeared perfectly chipper for a day or so afterwards, in forty-eight hours their hair was on end, their breath began to come in little hiccups. In five days they were dead, with exactly those symptoms their brothers had, after injections of the living diphtheria bacilli. So it was that Emile Roux discovered the diphtheria poison. . . .

By itself this weird experiment of the gigantic dose of feebly poisonous soup would only have made microbe hunters laugh. It was scandalous. "What!—if a great flask of diphtheria microbes can make so little poison that it takes a good part of a bottle of it to kill a small guinea-pig—how can a few microbes in a child's throat make enough to do that child to death? It is idiotic!"

But Roux had got his start. With this silly experiment as an uncertain flashlight, he went tripping and stumbling through the thickets, he bent his sallow bearded face (sometimes it was like the face of some unearthly bird of prey) over a precise long series of tests. Then suddenly he was out in the open. Presently, it was not more than two months later, he hit on the reason his poison had been so weak before—he simply hadn't left his germ-filled bottles in the incubator for long enough; there hadn't been time enough for them really to get down to work to make their deadly stuff. So, instead of four days, he left the microbes stewing at body temperature in their soup for forty-two days, and when he ran that brew through the filter—presto! With bright eyes he watched unbelievably tiny amounts of it do dreadful things to his animals—he couldn't seem to cut down the dose to an amount small enough to keep it from doing sad damage to his guinea-pigs. Exultant he watched feeble drops of it do away with rabbits, murder sheep, lay large dogs low. He played with this fatal fluid; he dried it; he tried to get at the chemistry of it (but failed); he got out a very concentrated essence of it though, and weighed it, and made long calculations.

One ounce of that purified stuff was enough to kill six hundred thousand guinea-pigs—or seventy-five thousand large dogs! And the bodies of those guinea-pigs who had got a six hundred thousandth of an ounce of this pure toxin—the tissues of those bodies looked like the sad tissues of a baby dead of diphtheria. . . .

So it was Roux made Loeffler's prophecy come true; it was

that way he discovered the fluid messenger of death which trickles from the insignificant bodies of diphtheria bacilli. But he stuck here; he had explained how a diphtheria germ murders babies but he had found no way to stop its maraudings. There was that letter from the mother—but Roux's researches petered out into various directions to doctors how to grow germs pure out of children's throats at the bedside, and into suggestions for useful gargles. . . . He hadn't Pasteur's tremendous grim stick-to-itiveness, nor his resourceful brain.

3

But away in Berlin there toiled another Emile—the Germans leave off the last "e"—Emil August Behring. He worked in Koch's laboratory, in the dilapidated building called the "Triangel" in the Schumann street. Here great things were stirring. Koch was there, no longer plain Doctor Koch of Wollstein, but now a Herr Professor, an eminent Privy Councilor. But his hat still fitted him; he peered through his spectacles, saying little; he was enormously respected, and against his own judgment he was trying to convince himself he had discovered a cure for tuberculosis. The authorities (scientists have reason occasionally to curse all authorities no matter how benevolent) were putting pressure on him. At least so it is whispered now by veteran microbe hunters who were there and remember those brave times.

"We have showered you with medals and microscopes and guinea-pigs—take a chance now, and give us a big cure, for the glory of the Fatherland, as Pasteur has done for the glory of France!" It was ominous stuff like this Koch was always hearing. He listened at last, and who can blame him, for what man can remain at his proper business of finding out the ways of microbes with Governments bawling for a place in the sun—or with mothers calling? So Koch listened and prepared his own disaster by telling the world about his "Tuberculin." But at the same time he guided his youngsters in fine jobs they were doing—and among these young men was Emil August

Behring. How Koch pointed the gun of his cold marvelous criticism at that poet's searchings!

And what a house of microbe hunters it was, that dingy Triangel! Its walls shook under the arguments and guttural cries and incessant experiments of Koch's young men. Paul Ehrlich was there, smoking myriads of cigars, smearing his clothes and his hands and even his face with a prismatic array of dyes, making bold experiments to find out how baby mice inherit immunity to certain vegetable poisons from their mothers. . . . Kitasato, the round-faced Japanese, was shooting lock-jaw bacilli into the tails of mice and solemnly amputating these infected tails—to see whether the creatures would perish from the poisons the microbes had made while the tails were still attached. . . . And there were many others there, some forgotten and some whose names are now famous. With a vengeance the Germans were setting out to beat the French, to bury them under a vast confusion of experiments, to save mankind first.

But particularly, Emil Behring was there. He was a little over thirty; he was an army doctor; he had a little beard, neater than Koch's scraggly one, but with less signs of originality. Just the same Behring's head, in spite of that prosaic beard, was the head of a poet; and yet, though he was fond of rhetoric, no one stuck closer to his laboratory bench than Behring. He compared the grandeur of the Master's discovery of the tubercle bacillus to the rosy tip of the snow-capped peak of his favorite mountain in Switzerland, while he probed by careful experiments into why animals are immune to microbes. He compared the stormy course of human pneumonia to the rushing of a mountain stream, while he discovered a something in the blood of rats—this stuff would kill anthrax bacilli! He had two scientific obsessions, which were also poetical: one was that blood is the most marvelous of the juices circulating in living things (what an extraordinary mysterious sap it was, this blood!)—the other was the strange notion (not a new one) that there must exist chemicals to wipe invading microbes out of animals and men—without hurting the men or the animals.

"I will find a chemical to cure diphtheria!" he cried, and inoculated herds of guinea-pigs with cultivations of virulent diphtheria bacilli. They got sick, and as they got sicker he shot various chemical compounds into them. He tried costly salts of gold, he tried naphthylamine, he tested more than thirty different strange or common substances. He believed innocently because these things could kill microbes in a glass tube without damaging the tube, they would also hit the diphtheria bacilli under a guinea-pig's hide without ruining the guinea-pig. But alas, from the slaughter house of dead and dying guinea-pigs his laboratory was, you would suppose he would have seen there was little to choose between the deadly microbes and his equally murderous cures. . . . Nevertheless, being a poet, Behring did not have too great a reverence for facts; the hecatombs of corpses went on piling up, but they failed to shake his faith in some marvelous unknown remedy for diphtheria hidden somewhere among the endless rows of chemicals in existence. Then, in his enthusiastic—but random—search he came upon the tri-chloride of iodine.

Under the skins of several guinea-pigs he shot a dose of diphtheria bacilli sure to kill them. In a few hours these microbes began their work; the spot of the injection became swollen, got ominously hot, the beasts began to droop—then, six hours after the fatal dose of the bacilli, Behring shot in his iodine tri-chloride. . . . "It is no good, once more," he muttered. The day passed with no improvement and the next morning the beasts began to go into collapses. Solemnly he put the guinea-pigs on their backs, then poked them with his finger to see if they could still scramble back on their feet. . . . "If the guinea-pig can still get up when you poke him, there may be yet a chance for him," explained Behring to his amazed assistants. What a test that was—think of a doctor having a test like this to see whether or no his patient would live! And what an abominably crude test! Less and less the iodine-treated guinea-pigs moved when he poked them— there was now no longer any hope. . . .

Then one morning Behring came into his laboratory to see

those guinea-pigs on their feet! Staggering about, and dreadfully scraggly looking beasts they were, but they were getting better from diphtheria, these creatures whose untreated companions had died days before. . . .

"I have cured diphtheria!" whispered Behring.

In a fever he went at trying to cure more guinea-pigs with this iodine stuff; sometimes the diphtheria bacilli killed these poor beasts; sometimes the cure killed them; once in a while one or two of them survived and crawled painfully back to their feet. There was little certainty of this horrible cure and no rime or reason. The guinea-pigs who survived, probably wished they were dead, for while the tri-chloride was curing them it was burning nasty holes in their hides too—they squeaked pitifully when they bumped these gaping sores. It was an appalling business!

Just the same, here were a few guinea-pigs, sure—except for this iodine—to have died of diphtheria; and they were alive! I often ponder how terrible was the urge forcing men like Behring to try to cure disease—they were not searchers for truth, but rabid, experimenting healers rather; ready to kill an animal or even a child maybe with one disease to cure him of another. They stopped at nothing. . . . For, with no evidence save these few dilapidated guinea-pigs, with no other proof of the virtues of this blistering iodine tri-chloride, Behring proceeded to try it on babies sick with diphtheria.

And he reported: "I have not been encouraged by certain carefully instituted tests of iodine tri-chloride on children sick with diphtheria. . . ."

But here were still some of those feeble but cured guinea-pigs, and Behring clutched at some good his murderous gropings might do. The gods were kind to him. He pondered, and at last he asked himself: "Will these cured animals be immune to diphtheria now?" He took these creatures and shot an enormous dose of diphtheria bacilli into them. They stood it! They never turned a hair at millions of bacilli, enough to kill a dozen ordinary animals. They were immune!

Now Behring no longer trusted chemicals (think of the

beasts that had gone down to the incinerator!) but he still had his fixed notion that blood was the most marvelous of the saps coursing through living things. He worshiped blood; his imagination gave it unheard-of excellences and strange virtues. So—with more or less discomfort to his decrepit cured guinea-pigs—he sucked a little blood with a syringe out of an artery in their necks; he let the tubes holding this blood stand until clear straw-colored serum rose over the red part of the blood. With care he drew this serum off with a tiny pipet—he mixed the serum with a quantity of virulent diphtheria bacilli: "Surely there is something in the blood of these creatures to make them so immune to diphtheria," pondered Behring; "undoubtedly there is something in this serum to kill the diphtheria microbes. . . ."

He expected to see the germs shrivel up, to watch them die, but when he looked, through his microscope, he saw dancing masses of them—they were multiplying, "exuberantly multiplying," he wrote in his notes with regret. But blood is wonderful stuff. Some way it must be at the bottom of his guinea-pig's immunity. "After all," muttered Behring, "this Frenchman, Roux, has proved it isn't the diphtheria germ but the poison it makes—it is the poison kills animals, and children. . . . Maybe these iodine-cured guinea-pigs are immune to the poison too!"

He tried it. With sundry guttural gruntings, with a certain poetic sloppiness, Behring got ready a soup which held poison but had been freed of microbes. Huge doses of this stuff he pumped from a syringe under the hides of his decreasing number of desolate cured guinea-pigs. Again, they were immune! Their sores went on healing, they grew fat. The poison bothered them no more than had the bacilli which made it. Here was something entirely new in microbe hunting, something Roux maybe dreamed of but couldn't make come true. Pasteur had guarded sheep against anthrax, and children from the bites of mad dogs, but here was something incredible—Behring, giving guinea-pigs diphtheria and then nearly killing them with his frightful cure, had made them proof against the

microbe's murderous toxin. He had made them immune to
the stuff of which one ounce was enough to kill seventy-five
thousand big dogs. . . .

"Surely it is in the blood I will find this antidote which
protects the creatures!" cried Behring.

He must get some of their blood. There were hardly any
of the battered but diphtheria-proof guinea-pigs left now, but
he must have blood! He took one of the veterans, and cut into
its neck to find the artery; there was no artery left—his numerous blood lettings had obliterated it. He poked about (let
us honor this animal!) and finally got a driblet of blood out of
a vessel in its leg. What a nervous time it was for Behring,
and I do not know whether it is Behring or his beasts who is
most to be pitied, for every morning he came down to the
laboratory wondering whether any of his priceless animals
were left alive. . . . But he had a few drops of serum now, from
a cured guinea-pig. He mixed this, in a glass tube, with a large
amount of the poisonous soup in which the diphtheria microbes had grown.

Into new, non-immune guinea-pigs went this mixture—*and
they did not die!*

"How true are the words of Goethe!" cried Behring.
"Blood is an entirely wonderful sap!"

Then, with Koch the master blinking at him, and with the
entire small band of maniacs in the laboratory breathless for
the result, Behring made his famous critical experiment. He
mixed diphtheria poison with the serum of a healthy guinea-
pig who was *not* immune, who had never had diphtheria or
been cured from it either, and this serum did not hinder one
bit the murderous action of the poison. He shot this mixture
into new guinea-pigs; in three days they grew cold; when he
laid them on their backs and poked them with his finger they
did not budge. In a few hours they had coughed their last sad
hiccup and passed beyond. . . .

"It is only the serum of immune animals—of beasts who
have had diphtheria and have been cured of it—it is only such
serum kills the diphtheria poison!" cried Behring. Healer that

he was, you can hear him muttering: "Now, maybe, I can make larger animals immune too, and get big batches of their poison-killing serum, then I'll try that on children with diphtheria . . . what saves guinea-pigs should cure babies!"

By this time nothing could discourage Behring. Like some victorious general swept on by the momentum of his first bloody success, he began shooting diphtheria microbes, and iodine tri-chloride, and the poison of diphtheria microbes, into rabbits, into sheep, into dogs. He tried to turn their living bodies into factories for making the healing serum, the toxin-killing serum. "Antitoxin" he called such serum. And he succeeded, after those maimings and holocausts and mistakes, always the necessary preludes to his triumphs. In a little while he had sheep powerfully immune, and from them he got plenty of blood. "Surely the antitoxin [he hadn't the faintest notion what the chemistry of this mysterious stuff was] certainly it will prevent diphtheria," said Behring.

He injected little doses of the sheep serum into guinea-pigs; the next day he pumped virulent diphtheria bacilli into these same beasts. It was marvelous to watch them. There they were, scampering about with never a sign of sickness, while their companions (who had got no protecting dose of serum) perished miserably in a couple of days. How good it was to see them die, those unguarded beasts! For it was these creatures told him how well the serum saved the other ones. Hundreds of pretty experiments of this kind Behring made (there was little sloppiness now) and his helpers maybe pointed to their foreheads, asking whether their chief would ever have done saving one set of guinea-pigs and killing another set to prove he had saved the first. But Behring had reasons. "We made so many experiments because we wanted to show Herr Koch how far we had come in our immunizing of laboratory animals," he wrote in one of his early reports.

There was only one fly in the ointment of his success—the guarding action of the antitoxin serum didn't last long. For a few days after guinea-pigs had got their injections of serum they stood big doses of the poison, but presently, in a week or two weeks, it took less and less of the toxin to kill them. Behring pulled at his beard: "This isn't practical," he muttered, "you couldn't go around giving all the children of Germany a shot of sheep serum every few weeks!" And alas, his eagerness for something to make the authorities wide-eyed, led him away from his fine fussings with a way to prevent diphtheria—it sent him a-whoring after the pound of cure. . . .

"Iodine tri-chloride is almost as bad for guinea-pigs as the microbes are—but this antitoxin serum, it doesn't give them sores and ulcers . . . I know it won't hurt my animals . . . I know it kills poison . . . now, if it would cure!"

Carefully he shot fatal doses of diphtheria bacilli into a lot of guinea-pigs. Next day, they were seedy. The second day their breath came anxiously. They stayed on their backs with that fatal laziness. . . . Then Behring took half of this lot of dying beasts, and into their bellies he injected a good heavy dose of the antitoxin from his immune sheep. Miracles! Nearly every one of them (but not all) began to breathe more easily in a little while. Next day, when he put them on their backs, they hopped nimbly back to their feet. They stayed there. By the fourth day they were as good as new, while their untreated companions, cold, dead, were being carried out by the animal boy. . . . The serum cured!

The old laboratory of the Triangel was in a furor now, over this triumphant finish of Behring's sloppy stumbling Odyssey. The hopes of everybody were purple—surely now he would save children! While he was getting ready his serum for the first fateful test on some baby near to death with diphtheria, Behring sat down to write his classic report on how he could cure beasts sure to die, by shooting into them a new, an unbelievable stuff their brother beasts had made in their own bodies—at the risk of nearly dying themselves. "We have no certain recipe for making animals immune," wrote Behring; "these experiments I have recorded do not include only my successes." Surely they did not, for Behring set down the messings and the fiascoes along with the few lucky stabs that gave him his sanguinary victory. . . . How *could* this pottering poet have pulled off the discovery of the diphtheria antitoxin? But then, come to think of it, those first ancient nameless men who invented sails to carry swift boats across the water—they must have groped that way too. . . . How many of the crazy craft of those anonymous geniuses turned turtle? It is the way discoveries are made. . . .

Toward the end of the year 1891, babies lay dying of diphtheria in the Bergmann clinic in the Brick Street in Berlin. On the night of Christmas, a child desperately sick with diphtheria cried and kicked a little as the needle of the first syringe full of antitoxin slid under its tender skin.

The results seemed miraculous. A few children died; the little son of a famous physician of Berlin passed out mysteriously a few minutes after the serum went into him and there was a great hullabaloo about that—but presently large chemical factories in Germany took up the making of the antitoxin in herds of sheep. Within three years twenty thousand babies had been injected and like a rumor spread the news, and Biggs, the eminent American Health Officer, then in Europe, was carried away by the excitement. He cabled dramatically and authoritatively to Dr. Park in New York:

DIPHTHERIA ANTITOXIN IS A SUCCESS; BEGIN TO PRODUCE IT.

In the excitement of this cure, those sad ones, who had lost dear ones through the first enthusiasm about the dangerous injections of the consumption cure of Koch, forgot their sorrow and forgave Koch because of his brilliant pupil Behring.

4

But there were still criticisms and muttered complaints, and this was natural, for the serum was no surefire, one hundred per cent curative stuff for babies—any more than it was for guinea-pigs. Then too, learned doctors pointed out that what happened under the hide of a guinea-pig was not the same—necessarily—as the savage thing going on in the throat of a

child. Thousands of children were getting the diphtheria serum, but some children (maybe not so many as before, perhaps?) kept dying horribly in spite of it. Doctors questioned.
. . . Some parents had their hopes dashed. . . .

Then Emile Roux came back into the battle. He discovered
brilliantly an easy way to make horses immune to the poison
—they did not die, they developed no horrid abscesses, and,
best of all, they furnished great gallon bottles full of the precious antitoxin—powerful stuff this serum was; little bits of it
destroyed large doses of that poison fatal to so many big dogs.

Like Behring—perhaps he was even more passionately sure
than Behring—Roux believed in advance this antitoxin would
save suffering children from death. He thought nothing of
prevention, he forgot about his gargles. He hurried to and fro
between his workroom and the stables, carrying big-bellied
flasks, jabbing needles into those patient horses' necks. Just
then, a particularly virulent breed (so Roux thought) of diphtheria bacillus was crawling through the homes of Paris. At the
Hospital for Sick Children, fifty out of every hundred children
(at least the statistics said so) were being carried blue-faced to
the morgue. At the Hospital Trousseau as many as sixty out
of a hundred were dying (but it is not clear whether the doctors there knew all these deaths to be from diphtheria). On
the first of February, 1894, Roux of the narrow chest and
hatchet face and black skull cap, walked into the diphtheria
ward of the Hospital for Sick Children, carrying bottles of his
straw-colored, miracle-working stuff.

In his study in the Institute in the Rue Dutot with a gleam
in his eye that made his dear ones forget he was marked for
death, there sat a palsied man, who must know, before he died,
whether one of his boys had wiped out another pestilence.
Pasteur waited for news from Roux. . . . Then too, all over
Paris there were fathers and mothers of stricken ones, praying
for Roux to hurry—they had heard of this marvelous cure of
Doctor Behring. It could almost bring babies back to life, folks
said—and Roux could see these people holding out their hands
to him. . . .

He got ready his syringes and bottles with the same cold steadiness the farmers had marveled at, long before, in those great days of the anthrax vaccine tests at Pouilly-le-Fort. His assistants, Martin and Chaillou, lighted the little alcohol lamp and hurried to anticipate his slightest order. Roux looked at the helpless doctors, then at the little lead-colored faces and the hands that picked and clutched at the edges of the covers, the bodies twisting to get a little breath. . . .

Roux looked at his syringes—did this serum really save life?

"Yes!" shouted Emile Roux, the human being.

"I don't know—let us make an experiment," whispered Emile Roux, the searcher for truth.

"But, to make an experiment, you will have to withhold the serum from half at least of these children—you may not do that." So said Emile Roux, the man with a heart, and all voices of all despairing parents were joined to the pleading voice of this Emile Roux.

"True, it is a terrible burden," answered the searcher that was Roux, "but just because this serum has cured rabbits, I do not *know* it will cure babies. . . . And I must know. I must find truth. Only by comparing the number of children who die, not having been given this serum, with the number who perish, having received it—only so can I ever know."

"But if you find out the serum is good, if it turns out from your experiment that the serum really cures—think of your responsibility for the death of those children, those hundreds of babies who did not get the antitoxin!"

It was a dreadful choice. There was one more argument the searcher that was Roux could have brought against the man of sentiment, for he might have asked: "If we do not find out surely, by experiment on these babies, the world may be lulled into the belief it has a perfect remedy for diphtheria—microbe hunters will stop looking for a remedy, and in the years that follow, thousands of children will die who might have been saved if hard scientific searching had gone on. . . ."

That would have been the final, the true answer of science to sentiment. But it was not made, and who after all can blame the pitying human heart of Roux for leaving the cruel road that leads to truth? The syringes were ready, the serum welled up into them as he gave a strong pull at the plungers. He began his merciful and maybe life-saving injections, and *every one* of the more than three hundred threatened children who came into the hospital during the next five months received good doses of the diphtheria antitoxin. Praise be, the results were a great vindication for the human Roux, for that summer, the experiment over, he told a congress of eminent medical men and savants from all parts of the world:

"The general condition of the children receiving the serum improves rapidly . . . in the wards there are to be seen hardly any more faces pale and lead-blue . . . instead, the demeanor of the children is lively and gay!"

He went on to tell the Congress of Buda-Pesth how the serum chased away the slimy gray membrane—that breeding place where the bacilli made their terrible poison—out of the babies' throats. He related how their fevers were cooled by this marvelous serum (it was like some breeze blowing from a lake of northern water across the fiery pavements of a city). The most dignified congress of prominent and celebrated physicians cheered. It rose to its feet. . . .

And yet—and yet—twenty-six out of every hundred babies Roux had treated—died, in spite of this marvelous serum. . . .

But it was an emotional time, remember, and Roux and the Congress of Buda-Pesth were not assembled to serve truth but to discuss and to plan and to celebrate the saving of lives. They cared little for figures then; they cared less for annoying objectors who carped about comparing figures; they were swept away by Roux's report of how the serum cooled fevered brows. Then, Roux could have answered such annoying critics (with the applause of his famous audience): "What if twenty-six out of a hundred did die—you must remember that for years before this treatment *fifty* out of a hundred died!"

And yet—I, who believe in this antitoxin, I say this, twenty years after—diphtheria is a disease having strange ups and downs of viciousness. In some terrible decades it kills its sixty out of a hundred; then some mysterious thing happens and the virus seems to weaken and only ten children are taken where sixty died before. So it was, in those brave days of Roux and Behring, for in a certain hospital in England, in those very days, the death rate from diphtheria had gone down from forty in a hundred to twenty-nine in a hundred—before the serum was ever used!

But the doctors at Buda-Pesth did not think of figures and they carried home the tidings of the antitoxin to all corners of the world, in a few years the antitoxin treatment of diphtheria became orthodox, and now there is not one doctor out of a thousand who will not swear that this antitoxin is a beautiful cure. Probably they are right. Indeed, there is evidence that when antitoxin is given on the first day of the disease, all but a few babies are saved—and if there is delay, many are lost. . . . Surely, any doctor should be called guilty, in the light of what is known, who did not give the antitoxin to a threatened child. I would be quick to call a doctor to give it to one of my own children. Why not, indeed? Perhaps the antitoxin cures. But it is not completely proved, and it is too late now to prove it one way or another to the hilt, because, since all the world believes in the antitoxin, no man can be found heartless enough or bold enough to do the experiment which science demands.

Meanwhile the searchers, believing, are busy with other things—and I can only hope, if another wave of the dreadful diphtheria of the eighties sweeps over the world again, I can only hope that Roux was right.

But even if the diphtheria antitoxin is not a sure cure, we already know that the experiments of Roux and Behring have not been in vain. It is a story still too recent, too much in the newspapers to be a part of this history—but to-day, in New York under the superb leadership of Dr. Park, and all over America, and in Germany, hundreds of thousands of babies and school-children are being ingeniously and safely turned into so many small factories for the making of antitoxin, so that they will never get diphtheria at all. Under the skins of these youngsters go wee doses of that terrible poison fatal to so many big dogs—but it is a poison fantastically changed so that it is harmless to a week-old baby!

There is every hope, if fathers and mothers can only be convinced and allow their children to undergo three small safe pricks of a syringe needle, that diphtheria will no longer be the murderer that it has been for ages.

And for this men will thank those first crude searchings of Loeffler and Roux and Behring.

7

Metchnikoff

The Nice Phagocytes

1

Microbe hunting has always been a queer humpty-dumpty business.

A janitor with no proper education was the first man to see microbes; a chemist put them on the map and made people properly afraid of them; a country doctor turned the hunting of them into something that came near to being a science; to save the lives of babies from the poison of one of the deadliest of them, a Frenchman and a German had to pile up mountains of butchered guinea-pigs and rabbits. Microbe hunting is a story of amazing stupidities, fine intuitions, insane paradoxes. If that is the history of the hunting of microbes, it is the same with the story of the science, still in its babyhood, of why we are immune to microbes. For Metchnikoff, the always excited searcher who in a manner of speaking founded that science—this Metchnikoff was not a sober scientific investigator; he was more like some hysterical character out of one of Dostoevski's novels.

Elie Metchnikoff was a Jew, and was born in southern Russia in 1845, and before he was twenty years old, he said: "I have zeal and ability, I am naturally talented—I am ambitious to become a distinguished investigator!"

He went to the University of Kharkoff, borrowed the then rare microscope from one of his professors, and after peering (more or less dimly) through it, this ambitious young man sat himself down and wrote long scientific papers before he had any idea at all of what science was. He bolted his classes for months on end, not to play, but to read; not to read novels mind you but to wallow through learned works on the "crystals of Proteic Substances" and to become passionate about inflammatory pamphlets whose discovery by the police would have sent him to the mines in Siberia. He sat up nights, drinking gallons of tea and haranguing his young colleagues (all of them forefathers of the present Bolsheviki) on atheism until they nicknamed him "God-Is-Not." Then, a few days before the end of the term, he crammed up the neglected lessons of months; and his prodigious memory, which was more like some weird phonograph record than any human brain, made it possible for him to write home to his folks that he had passed first and got a gold medal.

Metchnikoff was always trying to get ahead of himself. He sent papers to scientific journals while he was still in his teens; he wrote these papers frantically a few hours after he had trained his microscope on some bug or beetle; the next day he would look at them again, and find that what he had been so certain of was not quite the same now. Hastily he wrote to the editor of the scientific journal: "Please do not publish the manuscript I sent you yesterday. I find I have made a mistake." At other times he was furious because his enthusiastic discoveries were turned down by the editors. "The world does not appreciate me!" he cried, and he went to his room, ready to die, dolefully whistling: "Were I small as a snail, I would hide myself in my shell."

But if Metchnikoff sobbed because his vivid talents were underestimated by his professors, he was also irrepressible. He forgot his contemplated suicides and his violent headaches in his incessant interest in all living things, but he was constantly spoiling his chances to do a good steady piece of scientific work by getting into quarrels with his teachers. Finally he told his mother (who had always spoiled him and believed in him): "I am especially interested in the study of protoplasm . . . but there is no science in Russia," so he rushed off to the University of Würzburg in Germany, only to find that he had arrived there six weeks ahead of the opening of school. He sought out some Russian students there, but they gave him the cold shoulder—he was a Jew—then, tired of life, he started back home, thinking of killing himself but with a few books in his satchel—and one of these was the just-published "Origin of Species" of Darwin. He read it, he swallowed the Theory of Organic Evolution with one great mental gulp, he became a bigoted supporter of it—from then on evolution was his religion until he began founding new scientific religions of his own.

He forgot his plans for suicide; he planned strange evolutionary researches; he lay awake nights, seeing visions—huge panoramas they were, of all beasts from cockroaches to elephants, as the children of some one remote and infinitely tiny ancestor. . . .

That conversion was Metchnikoff's real start in life, for now he set out (and kept at it for ten years), quarreling and expostulating his way from one laboratory to another, from Russia through Germany to Italy, and from Italy to the island of Heligoland. He worked at the evolution of worms. He accused the distinguished German zoölogist Leuckart of stealing his stuff; incurably clumsy with his fingers, he clawed desperately into a lizard to find the story of evolution its insides might tell him—and when he could not find what he wanted, he threw what was left of the reptile across the laboratory. Unlike Koch or Leeuwenhoek, who were great because they knew how to ask questions of nature, Metchnikoff read books on Evolution, was inspired, shouted "Yes!" and then by vast sloppy experiments proceeded to try to force his beliefs down nature's throat. Strange to say, sometimes he was right, importantly right as you will see. Up till now (it was in the late eighteen seventies) he knew nothing about microbes, but all the time his mania to prove the survival of the fittest was driving him toward his fantastic theory—partly true—of how mankind resists the assaults of germs.

Metchnikoff's first thirty-five years were a hubbub and a perilously near disastrous groping toward this event—toward that great notoriety that waited for him on the Island of Sicily in the Mediterranean Sea. At twenty-three he had married Ludmilla Feodorovitch, who was a consumptive and had to be carried to the wedding in an invalid's chair. Then followed a pitiful four years for them. They dragged about Europe, looking for a cure; Metchnikoff trying in odd moments snatched from an irritatedly tender nursing of his wife, to do experiments on the development of green flies and sponges and worms and scorpions—trying above all to make some sensational discovery which might land him a well-paid professorship. "The survivors are not the best but the most cunning," he whispered, as he published his scientific papers and pulled his wires. . . .

Finally Ludmilla died; she had spent her last days solaced by morphine, and now Metchnikoff, who had caught the habit from her, wandered from her grave through Spain to Geneva, taking larger and larger doses of the drug—meanwhile, his eyes hurt him terribly, and what is a naturalist, a searcher, without eyes?

"Why live?" he cried, and took a dose of morphine that he knew must kill him, but the dose was too large, he became nauseated and threw it up. "Why live?" he cried again and took a hot bath and rushed out in the open air right afterwards to try to catch his death of pneumonia. But it seems that the wise witty gods who fashion searchers had other purposes for him. That very night he stopped, agape at the spectacle of a cloud of insects swirling round the flame of a lantern. "These insects live only a few hours!" he cried to himself. "How can the theory of the survival of the fittest be applied to them?" So he plunged back into his experiments.

Metchnikoff's grief was terrific but it did not last long. He was appointed Professor at the University of Odessa, and there he taught the Survival of the Fittest and became respected for his learning, and grew in dignity, and in less than two years after the death of Ludmilla, he had met Olga, a bright girl of fifteen, the daughter of a man of property. "His appearance is not unlike that of the Christ—he is so pale and seems so sad," whispered Olga. Soon after they were married.

From then on Metchnikoff's life was much less disastrous; he tried far less often to commit suicide; his hands began to catch up with his precocious brain—he was learning to do experiments. Never was there a man who tried more sincerely to apply his religion (which was science) to every part of his life. He took Olga in hand and taught her science and art, and even the art and science of marriage! She worshipped the profound certainties that science gave him, but said, long afterwards: "The scientific methods which Metchnikoff applied to everything might have been a grave mistake at this delicate psychological moment. . . ."

2

It was in 1883, when the discoveries of Pasteur and Koch had made everybody mad about microbes, that Metchnikoff turned suddenly from a naturalist into a microbe hunter. He had wrangled with the authorities of the University of Odessa, and departed for the Island of Sicily with Olga and her crowd of little brothers and sisters, and here he set up his amateur laboratory in the parlor of their cottage looking across the magic water to the blue Calabrian shore. His intuition told him that microbes were now the thing in science and he dreamed about making great discoveries of new microbes—he was sincerely interested in them as well, but he knew nothing about the subtle ways of hunting them, indeed he had hardly seen a germ. He stamped about his parlor-laboratory, expounding biological theories to Olga, studying starfish and sponges, telling the children fairy stories, doing everything in short that was as far as possible removed from those thrilling researches of Koch and Pasteur. . . .

Then, one day, he began to study the way sponges and starfishes digest their food. Long before he had spied out strange cells inside these beasts, cells that were a part of their bodies, but cells that were free-lances, as it were, moving from place to place through the carcasses of which they formed a part, sticking out one part of themselves and dragging the rest of themselves after the part they had stuck out. Such were the *wandering cells*, which moved by flowing, exactly like that small animal, the ameba.

Metchnikoff sat down before his parlor table, and with that impatient clumsiness of a man whose hands seem unable to obey his brain, he got some little particles of carmine into the insides of the larva of a starfish. This was an ingenious and very original trick of Metchnikoff's, because these larvae are as transparent as a good glass window; so he could see, through his lens, what went on inside the beast; and with ex-cited delight he watched the crawling, flowing free-lance cells in this starfish ooze toward his carmine particles—and eat them up! Metchnikoff still imagined he was studying the di-gestion of his starfish, but strange thoughts—that had nothing to do with such a commonplace thing as digestion—little fog-wraiths of new ideas began to flutter through his head. . . .

The next day Olga took the children to the circus to see some extraordinary performing monkeys. Metchnikoff sat alone in his parlor, tugging at his biblical beard, gazing with-out seeing them at his bowls of starfish. Then—it was like that blinding light that bowled Paul over on his way to Damascus—in one moment, in the most fantastical, you would say impossible flash of a second, Metchnikoff changed his whole career.

"These wandering cells in the body of the larva of a starfish, these cells eat food, they gobble up carmine granules—but they must eat up microbes too! Of course—the wandering cells are what protect the starfish from microbes! *Our* wan-dering cells, the white cells of our blood—they must be what protects us from invading germs . . . they are the cause of immunity to diseases . . . they are what keep the human race from being killed off by malignant bacilli!"

Without one single bit of evidence, without any research at all, Metchnikoff jumped from the digestions of starfish to the ills of men. . . .

"I suddenly became a pathologist," he wrote in his diary (and this was not much more strange than if a cornet player should suddenly announce himself as an astrophysicist!) ". . . Feeling that there was in this idea something of surpassing interest, I became so excited that I began striding up and down the room, and even went to the seashore to collect my thoughts."

Now Koch, precise microbe hunter that he was, would hardly have trusted Metchnikoff with the wiping of his microscope, but his ignorance of germs was nothing to this wild Russian.

"I said to myself that, if my theory was true, a sliver put into the body of a starfish larva . . . should soon be surrounded by wandering cells. . . ." And he remembered that when men run splinters into their fingers, and neglect to pull them out, those splinters are soon surrounded by pus—which consists largely of the wandering white cells of the blood. He rushed out into the garden back of the cottage, pulled some rose thorns off a little shrub which he had decorated as a Christmas tree for Olga's brothers and sisters; he dashed back into his absurd laboratory and stuck these thorns into the body of one of his water-clear young starfish. . . .

Up he got, at dawn the next morning, full of wild hopes, —and he found his guess had come true. Around the rose-slivers in the starfish were sluggish crawling masses of its wandering cells! Nothing more was necessary (such a jumper at conclusions was he) to stamp into his brain the fixed idea that he now had the explanation of all immunity to diseases; he rushed out that morning to tell famous European professors, who happened then to be in Messina, all about his great idea. "Here is why animals can withstand the attacks of microbes," he said, and he talked with such enthusiastic eloquence about how the wandering cells of the starfish tried to eat the rose thorns (and he could show it so prettily too) that even the most eminent and pope-like Professor Doctor Virchow (who had sniffed at Koch) believed him!

Metchnikoff was now a microbe hunter. . . .

3

With Olga and the children flapping along and keeping up as best they could, Metchnikoff hurried to Vienna to proclaim his theory that we are immune to germs because our bodies have wandering cells to gobble germs up; he made a bee-line for the laboratory of his friend, Professor Claus—who was a zoölogist, and knew nothing about microbes either, and so was properly amazed:

"I would be greatly honored to have you publish your theory in my Journal," said Claus.

"But I must have a scientific name for these cells that devour microbes—a Greek name—what would be a Greek name for such cells?" cried Metchnikoff.

Claus and his learned colleagues scratched their heads and peered into their dictionaries and at last they told him: "Phagocytes! Phagocyte is Greek for devouring cell—phagocytes is what you must call them!"

Metchnikoff thanked them, tacked the word "phagocyte" to the head of his mast, and set sail on the seas of his exciting career as a microbe hunter with that word as a religion, an explanation of everything, a slogan, a means of gaining a living—and, though you may not believe it, that word did result in something of a start at finding out how it is we are immune! From then on he preached phagocytes, he defended their reputations, he did some real research on them, he made enemies about them, he doubtless helped to start the war of 1914 with them, by the bad feeling they caused between France and Germany.

He went from Vienna to Odessa, and there he gave a great

scientific speech on "The Curative Forces of the Organism" to the astonished doctors of the town. His delivery was superb; his sincerity was undoubted—but there is no record of whether or not he told the amazed doctors that he had not, up till then, so much as seen one phagocyte gobble up a single malignant microbe. Everybody—and this includes learned doctors—will stop to watch a dog fight; so this idea of Metchnikoff's, this story of our little white blood cells rushing to an endless series of Thermopylæs to man the pass against murderous germs—this yarn excited them, convinced them. . . .

But Metchnikoff knew he would have to have real evidence, and presently he found it, beautifully clear, in water fleas. For a time he forgot speeches and began fishing water fleas out of ponds and aquariums; here he was deucedly ingenious again, for these small animals, like starfish larvæ, were transparent so that he could see through his lens what went on inside them. For once he grew patient, and searched, like the real searcher that he so rarely was, for some disease that a water flea perchance might have. This history has already made it clear that microbe hunters usually find other things than they set out to look for—but Metchnikoff just now had different luck; he watched his water fleas in their aimless daily life, and suddenly, through his lens he saw one of these beasts swallow the sharp, needle-like spores of a dangerous yeast. Down into the wee gullet went these needles, through the walls of the flea's stomach they poked their sharp points, and into the tiny beast's body they glided. Then—how could the gods favor such a wild man so!—Metchnikoff saw the wandering cells of the water flea, the *phagocytes* of this creature, flow towards those perilous needles, surround them, eat them, melt them up, digest them. . . .

When—and this happened often too and so made his theory perfect—the phagocytes failed to go out to battle against the deadly yeast needles, these invaders budded rapidly into swarming yeasts, which in their turn ate the water flea, poisoned him—and that meant good-by to him!

Here Metchnikoff had peeped prettily into a thrilling, deadly struggle on a tiny scale, he had spied upon the up till now completely mysterious way in which *certain* living creatures defend themselves against their would-be assassins. His observations were true as steel, and you will have to grant they were devilishly ingenious, for who would have thought to look for the why of immunity in such an absurd beast as the water flea? Now Metchnikoff needed nothing more to convince him of the absolute and final rightness of his theory, he probed no deeper into this struggle (which Koch would have spent years over) but wrote a learned paper:

"The immunity of the water flea, due to the help of its phagocytes, is an example of natural immunity . . . for, once the wandering cells have not swallowed the yeast spore at the moment of its penetration into the body, the yeast germinates . . . secretes a poison which drives the phagocytes back not only, but kills them by dissolving them completely."

4

Then Metchnikoff went to see if this same battle took place in frogs and rabbits, and suddenly, in 1886, the Russian people were thrilled by Pasteur's saving of sixteen of their folk from the bite of the mad wolf. The good people of Odessa and the farmers of the Zemstvo round about gave thanks to God, hurrahs for Pasteur, and a mighty purse of roubles for a laboratory to be started at once in Odessa. And Metchnikoff .was appointed Scientific Director of the new Institute—for had not this man (they forgot for a moment he was Jewish) studied in all the Universities of Europe, and had he not lectured learnedly to the doctors of Odessa, telling about the phagocytes of the blood, which gobble microbes?

"Who knows?" you can hear the people saying. "Maybe in our new Institute, Professor Metchnikoff can train these little phagocytes to gobble up all microbes?"

Metchnikoff accepted the position, but told the authorities, shrewdly: "I am only a theoretician; I am overwhelmed with researches—some one else will have to be trained to make vaccines, to do the practical work."

Nobody in Odessa knew anything about microbe hunting then, so Metchnikoff's friend, Doctor Gamaléia, was sent to the Pasteur Institute in Paris posthaste. The citizens were anxious to begin to be prevented from having diseases; they bawled for vaccines. So Gamaléia, after a little while in Paris, where he watched Roux and Pasteur and learned a great deal from them, but not quite enough—this Gamaléia came back and started to make anthrax vaccines for the sheep of the Zemstvo, and rabies vaccines for the people of the town. "All should now go very well!" cried Metchnikoff (he knew nothing of the nasty tricks virulent microbes can play) and he retired to his theoretical fastnesses to grapple with rabbits and dogs and monkeys, to see if their phagocytes would swallow the microbes of consumption and relapsing fever and erysipelas. Scientific papers vomited from his laboratory, and the searchers of Europe began to be excited by the discoveries of this strange genius in the south of Russia. But he began to have troubles with his theory, for dogs and rabbits and monkeys—alas—are not transparent, like water fleas. . . .

Then the shambles began. Gamaléia and the other members of Metchnikoff's practical staff began to fight among themselves and mix up vaccines; microbes spilled out of tubes; the doctors of the town—naturally a little jealous of this new form of healing—started to snoop into the laboratory, to ask embarrassing questions, to start whispers going through the town: "Who is this Professor Metchnikoff—he hasn't even a doctor's certificate. He is only a naturalist, a mere bug-hunter—how can he know anything about preventing diseases?"

"Where are those cures?" demanded the people. "Give us our preventions!" shouted the farmers—who had gone down into their socks for good roubles. Metchnikoff came out of the fog of his theory of phagocytes for a moment, and tried to satisfy them by sowing chicken cholera bacilli among the meadow mice which were eating up the crops. But, alas, a lying, inflammatory report appeared in the daily paper, screaming that this Metchnikoff was sowing death—that chicken cholera could change into human cholera. . . .

"I am overwhelmed with my researches," muttered Metchnikoff. "I am a theoretician—my researches need a peaceful shelter in which to be developed. . . ." So he asked for a vacation, got it, packed his bag, and went to the Congress of Vienna to tell everybody about phagocytes, and to look for a quiet place in which to work. He *must* get away from that dreadful need to prove that his theories were true by dishing out cures to impatient authorities and peasants who insisted on getting their money's worth out of research. From Vienna he went to Paris to the Pasteur Institute, and there a great triumph and surprise waited for him. He was introduced to Pasteur, and at once Metchnikoff exploded into tremendous explanations of his theory of phagocytes. He made a veritable movie of the battle between the wandering cells and microbes. . . .

The old captain of the microbe hunters looked at Metchnikoff out of tired gray eyes that now and then sparkled a little: "I at once placed myself on your side, Professor Metchnikoff," said Pasteur, "for I have been struck by the struggle between the divers microörganisms which I have had occasion to observe. I believe you are on the right road."

Although the struggles Pasteur mentioned had nothing to do with phagocytes gobbling up microbes, Metchnikoff—and this is not unnatural—was filled with a proud joy. The greatest of all microbe hunters really understood him, believed in him. . . . Olga's father had died, leaving them a modest income, here in Paris his theory of phagocytes would have the prestige of a great Institute back of it. "Is there a place for me here?" he asked. "I wish only to work in one of your laboratories in an honorary capacity," begged Metchnikoff.

Pasteur knew how important it was to keep the plain people thrilled about microbe hunting—it is the *drama* of science that they can understand—so Pasteur said: "You may not only come to work in our laboratory, but you shall have an entire laboratory to yourself!" Metchnikoff went back to Odessa, getting a dreadful snubbing from Koch on the way, and wondered whether it would not be best to give up his tidy salary at the Russian Institute, to get away from these people yelling for results. . . . But he began to take up his work again, when suddenly something happened that left no doubt in his mind as to what he had better do.

In response to the farmers' complaints of "Where are your vaccines, our flocks are perishing from anthrax!" Metchnikoff had told Dr. Gamaléia to start giving sheep the anthrax vaccine on a large scale. Then, one bright morning, while the Director was with Olga in their summer home, in the country, a fearful telegram came to him from Gamaléia:

MANY THOUSANDS OF SHEEP KILLED BY THE ANTHRAX VACCINE.

A few months later they were safely installed in the new Pasteur Institute in Paris, and Olga (who enjoyed painting and sculpture much better—but who would do anything for her husband because he was a genius, and always kind to her) this good wife, Olga, held his animals and washed his bottles for Metchnikoff. From then on they marched, hand in hand, over a road strewn with their picturesque mistakes, from one triumph to always greater victories and notorieties.

5

Metchnikoff bounced into the austere Pasteur Institute and started a circus there which lasted for twenty years; it was as if a skilled proprietor of a medicine show had become pastor of a congregation of sober Quakers. He came to Paris and found himself already notorious. His theory of immunity—it would be better to call it an exciting romance, rather than a theory—this story that we are immune because of a kind of battle royal between our phagocytes and marauding microbes, this yarn had thrown the searchers of Europe into an uproar. The microbe hunters of Germany and Austria for the most part did not believe it—on the contrary, tempted to believe it by its simplicity and prettiness, they denied it with a peculiar violence. They denounced Metchnikoff in congresses and by experiments. One old German, Baumgarten, wrote a general denunciation of phagocytes, on principle, once a year, in an important scientific journal. For a little while Metchnikoff wavered; he nearly swooned, he couldn't sleep nights, he thought of going back to his soothing morphine; he even contemplated suicide once more—oh! why could not those nasty Germans see that he was right about phagocytes? Then he recovered. Something seemed to snap in his brain, he became courageous as a lion, he started a battle for his theory—it was a grotesque, partly scientific wrangle—but, in spite of all its silliness, it was an argument that laid the foundations of the little that is known to-day about why we are immune to microbes.

"I have demonstrated that the serum of rats kills anthrax germs—it is the *blood* of animals not their phagocytes that makes them immune to microbes," shouted Emil Behring, and all the bitter enemies of Metchnikoff sang Aye in the chorus. The scientific papers published to show that blood is the one important thing would fill three university libraries.

"It is the phagocytes that eat up germs and so defend us," roared Metchnikoff in reply. And he published ingenious experiments which proved anthrax bacilli grow exuberantly in the blood of sheep which have been made immune by Pasteur's vaccine.

Neither side would budge from this extreme, prejudiced position. For twenty years both sides were so enraged they could not stop to think that perhaps both our blood *and* our phagocytes might work together to guard us from germs. That fight was a kind of magnificent but undignified shouting of "You're a liar— On the contrary, it's you that's the liar!" which blinded Metchnikoff and his opponents to the idea that it might be neither the blood nor the phagocytes which are at the bottom of our resistance to some diseases. If they had only stopped for a moment, wiped their brows and cleaned the blood from their mental noses, to remember how little they knew, how slowly they should go—considering what subtle complicated stuff this blood and those phagocytes are—if they had only remembered how foolish, in the darkness of their ignorance, it was to cook up any explanation at all of why we are immune! If Metchnikoff had only kept on, obscure in Odessa, with his beautiful researches on the why of the wandering cells of the water fleas eating up those terrible little yeasts. . . . If he had only been patient and tried to get to the bottom of that!

But the stumbling strides of microbe hunters are not made by any perfect logic, and that is the reason I may write a grotesque, but not perfect story of their deeds.

In the grand days of Pasteur's fight with anthrax and his victory against rabies, he had worked like some subterranean distiller of secret poisons, with only Roux and Chamberland and one or two others to help him. In that dingy laboratory in the Rue d'Ulm he had been very impolite, even nasty, to all curious intruders and ambitious persons. He even chased adoring pretty ladies away. But Metchnikoff!

Here was an entirely different sort of searcher. Metchnikoff had an immensely impressive beard and a broad forehead that crowned eyes which squinted vividly—and intelligently—from behind his spectacles. His hair grew down over the back of his neck in a way that showed you he was too deep in thoughts to think of having it cut. He knew everything! He could tell —and it was authentic—of countless biological mysteries; he had seen the wandering cells of a tadpole turn it into a frog by eating the tadpole's tail, and he had built circles of fire around scorpions to show that these unhappy creatures, failing to find a way out, do not commit suicide by stinging themselves to death. He told these horrors in a way to make you feel the remorseless flowing and swallowing of the wandering cells—you could hear the hissing of the doomed and baffled scorpion. . . .

He had brilliant ideas for experiments and was always trying to carry out these ideas—intensely—but at any moment he was ready to drop his science to praise the operas of Mozart or whistle the symphonies of Beethoven, and sometimes he seemed to be more learned about the dramas and the loves of Goethe than about those phagocytes upon which his whole fame rested. He refused to wear a high hat toward lesser men; he would see any one and was ready to believe anything—he even tried the remedies of patent medicine quacks on dying guinea-pigs. And he was a kind man. When his friends were sick he overwhelmed them with delicacies and advice and shed sincere tears on their pillows—so that finally they nicknamed him "Mamma Metchnikoff." His views on the intimate instincts and necessities of life were astoundingly unlike those of any searcher I have ever heard of. "The truth is that artistic genius and perhaps all kinds of genius are closely associated with sexual activity . . . so, for example, an orator speaks better in the presence of a woman to whom he is devoted."

He insisted that he could experiment best when pretty girls were close by!

Metchnikoff's workshop in the Pasteur Institute was more than a mere laboratory; it was a studio, it had the variegated attractions of a country fair; it radiated the verve and gusto of a three-ringed circus. Is it any wonder, then, that young doctors, eager to learn to hunt microbes, flocked to him from all over Europe? Their brains responded to this great searcher who was also a hypnotist, and their fingers flew to perform the ten thousand experiments, ideas for which belched out of the mind of Metchnikoff like an incessant eruption of fireworks.

"Mr. Saltykoff!" he would cry. "This student of Professor Pfeiffer in Germany claims that the serum of a guinea-pig will keep other guinea-pigs from dying of hog-cholera. Will you be so good as to perform an experiment to see if that is so?" And the worshiping Saltykoff rushed off—knowing what the master wanted to prove—to show that the German claims were nonsense. For a hundred other intricate tests, for which his own fingers were too impatient, Metchnikoff called upon Blagovestchensky, or Hugenschmidt, or Wagner, or Gheorgiewski, or the now almost forgotten Sawtchenko. Or when these were all busy, then there was Olga to be lured away from her paints and clay models—Olga could be depended upon to prove the most delicate points. In that laboratory there were a hundred hearts that beat as one and a hundred minds with but a single thought—to write the epic of those tiny, roundish, colorless, wandering cells of our blood, those cells, which, smelling from afar the approach of a murderous microbe, swam up the current of the blood, crawled strangely through the walls of the blood vessels to do battle with the germs and so guard us from death.

The great medical congresses of those brave days were exciting debating societies about microbes, about immunity, and it was in the weeks before a congress (Metchnikoff always went to them) that his laboratory buzzed with an infernal rushing to and fro. "We must hurry," Metchnikoff exclaimed, "to make all of the experiments necessary to support my arguments!" The crowd of adoring assistants then slept two hours less each night; Metchnikoff rolled up his sleeves, too, and seized a syringe. Young rhinoceros beetles, green frogs, alligators, or weird Mexican axolotls were brought from the animal house by the sweating helpers (sometimes the ponds were dredged for perch and gudgeon). Then the mad philosopher, his eyes alight, his broad face so red that it glowed like some smoldering brush-fire under his beard, his mustaches full of bacilli spattered into it by his excited and poetic gestures—this Metchnikoff, I say, proceeded to inject swarms of microbes into one or another of his uncomplaining, cold-blooded menagerie. "I multiply experiments to support my theory of phagocytes!" he was wont to say.

6

It is amazing, when you remember that his brain was always inventing stories about nature, how often these stories turned out to be true when they were put to the test of experiment. A German hunter had claimed: "There is nothing to Metchnikoff's theory of phagocytes. Everybody knows that you can see microbes inside of phagocytes—they have undoubtedly been gobbled up by the phagocytes. But these wandering cells are not defenders, they are mere scavengers—they will only swallow dead microbes!" The London Congress of 1891 was drawing near; Metchnikoff shouted for some guinea-pigs, vaccinated them with some cholera-like bacilli that his old friend, the unfortunate Gamaléia, had discovered. Then, a week or so later, the big-bearded philosopher shot some of these living, dangerous bacilli into the bellies of vaccinated beasts. Every few minutes, during the next hours, he ran slender glass tubes into their abdomens, sucked out a few drops of the fluid there, and put it before the more or less dirty lens of his microscope, to see whether the phagocytes of the immune beasts were eating up Gamaléia's bacilli. Presto! These roundish crawling cells were crammed full of the microbes!

"Now I shall prove that these microbes inside the phagocytes are still alive!" cried Metchnikoff. He killed the guinea-pig, slashed it open, and sucked into another little glass tube some of the grayish slime of wandering cells which had gathered in the creature's belly to make meals off the microbes. In a little while—for they are very delicate when you try to keep them alive outside the body—the phagocytes had died, burst open, and the *live* bacilli they had swallowed galloped out of them! Promptly, when Metchnikoff injected them, these microbes that had been swallowed, murdered guinea-pigs who were not immune.

By dozens of brilliant experiments of this kind, Metchnikoff forced his opponents to admit that phagocytes, sometimes, can eat vicious microbes. But the pitiful waste of this brainy Metchnikoff's life was that he was always doing experiments to defend an idea, and not to find the hidden truths of nature. His experiments were weird, they were often fantastically entertaining, but they were so artificial—they were so far away from the point of what it is that makes us immune. You would think that his brain, which seemed to be able to hold all knowledge, would have dreamed of subtle tests to find out just how it is that one child can be exposed to consumption and never get it, while some carefully and hygienically raised young girl dies from consumption at twenty. *There* is the riddle of immunity (and it is still completely a riddle!). "Oh! it is doubtless due to the fact that her phagocytes are not working!" Metchnikoff would have exclaimed, and then he might rush off to flabbergast some opponent by proving that the phagocytes of an alligator eat up typhoid fever bacilli—which never bother alligators anyway.

The devotion of the workers in his laboratory was amazing. They let him feed them virulent cholera bacilli (even one of those pretty inspirational girls swallowed them!) to prove that the blood has nothing to do with our immunity to cholera. For years—he himself said that it was an insanity of his—he was fond of toying with the lives of his researching slaves, and the only thing that excused him was his perfect readiness to risk death along with them. He swallowed more tubes of cholera bacilli than any of them. In the midst of this dangerous business, one of the assistants, Jupille, became violently sick with real Asiatic cholera and Metchnikoff's remorse was immoderate. "I shall never survive the death of Jupille!" he moaned, and Olga, that good wife, had to be on her guard day and night to keep her famous husband from one of his (always fruitless) attempts at suicide. At the end of these strange experiments, Metchnikoff jabbed needles into the arms of the survivors, drew blood from them, and triumphantly found that this blood did not protect guinea-pigs from doses of virulent cholera germs. How he hated the idea of blood having any importance! "Human cholera gives us another example," he wrote, "of a malady whose cure cannot be explained by the preventive properties of the blood."

When some more than ordinarily independent student would come whispering to him that he had discovered a remarkable something about blood, Metchnikoff became magnificent like Moses coming down off Mt. Sinai—searchers for mere truth had a bad time in that laboratory, and you can imagine the great dauntless champion of phagocytes ordering a dissenter from his theory to be burned, and then weeping inconsolably over him afterwards. But, just the same,

Metchnikoff—so great was the number of experiments made by an always changing crowd of eager experimenters in his laboratory—this Metchnikoff was partly responsible for the discovery of some of the most astounding virtues of blood. For, in the midst of his triumphs, Jules Bordet came to work with the master. This Bordet was the son of the schoolmaster of the village of Soignies in Belgium. He was timid, he seemed insignificant, he had careless ways and water-blue, absent-minded eyes—eyes that saw things nobody else was looking for. Bordet set to work there, and right in the shadow of the master's beard, while the walls shook with the slogan "Phagocytes!"—the Belgian pried into the mystery of how blood kills germs; he laid the foundation for those astounding delicate tests which tell whether blood is human blood, in murder cases. It was here too, that Bordet began the work which led, years later, to the famous blood test for syphilis— the Wassermann reaction. Metchnikoff was often annoyed with Bordet, but he was proud of him too, and whenever Bordet found anything in blood that was harmful to microbes, and might help to make people immune to them, Metchnikoff consoled himself by inventing more or less accurate experiments which showed that these microbe-killing things came from the phagocytes, after all. Bordet did not remain long in Metchnikoff's laboratory. . . .

Toward the end of the nineteenth century, when romantic microbe hunting began to turn into a regular profession, recruited from good steady law-abiding young doctors who were not prophets or reckless searchers—in those days Metchnikoff's bitter trials with people who didn't believe him began to be less terrible. He received medals and prizes of money, and even the Germans clapped their hands and were respectful when he walked majestically into some congress. A thousand searchers had spied phagocytes in the act of gobbling harmful germs—and although that did not explain at all why one man dies from an attack of pneumonia microbes, while another breaks into a sweat and gets better—just the same there is no doubt that pneumonia germs are sometimes eaten and so got rid of by phagocytes. So Metchnikoff, after you discount his amazing illogic, his intolerance, his bullheadedness, really did discover a fact which may make life easier for suffering mankind. Because, some day, a dreamer, an experimenting genius like the absent-minded Bordet may come along—and he may solve the riddle of why phagocytes sometimes gobble germs and sometimes do not—he might even teach phagocytes always to eat them. . . .

7

At last Metchnikoff began really to be happy. His opponents were partly convinced, and partly they stopped arguing with him because they found it was no use—he could always experiment more tirelessly than they, he could talk longer, he could expostulate more loudly. So Metchnikoff, at the beginning of the twentieth century, sat down to write a great book on all that he had found out about why we are immune. It was an enormous treatise you would think it would take a lifetime to write. It was written in a style Flaubert might have envied. He made every one of the ten thousand facts in it vivid, and every one of them was twisted prettily to prove his point. It is a strange novel with a myriad of heroes—the wandering cells, the phagocytes of all the animals of the earth.

His fame made him take a real delight in being alive. Twenty years before, detesting the human race, sorry for himself, and hating life, he had told Olga: "It is a crime to have children—no human being should consciously reproduce himself." But now that he had begun to take delight in existence, the children of Sèvres, the suburb where he lived, called him "Grandpa Christmas" as he patted their heads and gave them candy. "Life is good!" he told himself. But how to hang onto it, now that it was slipping away so fast. In only one way, of course—by science!

"Disease is only an episode!" he wrote. "It is not enough to cure (he had discovered no cures) . . . it is necessary to find out what the destiny of man is, and why he must grow old and die when his desire to live is strongest." Then Metchnikoff abandoned work on his dead phagocytes and set out to found fantastic sciences to explain man's destiny, and to avoid it. To one of these, the science of old age, he gave the sonorous name "Gerontology," and he gave the name "Thanatology" to the science of death. What awful sciences they were; the ideas were optimistic; the observations he made in them were so inaccurate that old Leeuwenhoek would have turned over in his grave had he known about them; the experiments Metchnikoff made, to support these sciences, would have caused Pasteur to foam with indignation that he had ever welcomed this outlandish Russian to his laboratory. And yet—and yet—the way really to prevent one of the most hideous microbic diseases came out of them. . . .

Metchnikoff dreaded the idea of dying but knew that he and everybody else would have to—so he set out to devise a hope (there was not one particle of science in this) for an easy death. Somewhere in his vast hungry readings, he had run across the report of two old ladies who had become so old that they felt no more desire for life—they wanted to die, just as all of us want to go to sleep at the end of a hard day's work. "Ha!" cried Metchnikoff, "that shows that there is an instinct for death just as there is an instinct for sleep! The thing to do is to find a way to live long enough in good health until we shall really crave to die!"

Then he set out on a thorough search for more of such lucky old ladies, he visited old ladies' homes, he rushed about questioning old crones, with their teeth out, who were too deaf to hear him. He went all the way from Paris to Rouen to interview (on the strength of a newspaper rumor) a dame reported to be a hundred and six. But, alas, all of the oldsters he talked to were strong for life, he never found any one like the two legendary old ladies. Just the same he cried: "There is a death instinct!" Contrary facts never worried him.

He studied old age in animals; and people were always sending him gray-haired dogs and dilapidated ancient cats; he published a solemn research on why a superannuated parrot lived to be seventy. He owned an ancient he-turtle, who lived in his garden, and Metchnikoff was overjoyed when this venerable beast—at the great age of 86—mated with two lady turtles and became the father of broods of little turtles. He dreaded the passing of the delights of love, and exclaimed, remembering his turtle: "Senility is not so profoundly seated as we suppose!"

But to push back old age? What is at the bottom of it? A Scandinavian scientist, Edgren, had made a deep study of the hardening of the arteries—that was the cause of old age, suggested Edgren, and among the causes of the hardening of the arteries were the drinking of alcohol, syphilis, and certain other diseases.

"A man is as old as his arteries, that is true," muttered Metchnikoff, and he decided to study the riddle of how that loathsome disease hardens the arteries. It was in 1903. He had just received a prize of five thousand francs, and Roux—who, though so different, so much more the searcher, had always stuck by this wild Metchnikoff—Roux had got the grand Osiris prize of one hundred thousand francs. Never were there two men so different in their ways of doing science, but they were alike in caring little for money, and together they decided to use all of these francs—and thirty thousand more which Metchnikoff had wheedled out of some rich Russians—to study that venereal plague, to attempt to give it to apes, to try to discover its then mysterious virus, to prevent it, cure it if possible. And Metchnikoff wanted to study how syphilis hardened the arteries.

So they bought apes with this money. French governors in the Congo sent black boys to scour the jungles for them, and presently large rooms at the Pasteur Institute were a-chatter with chimpanzees and orang-outangs, and the cries of these were drowned out by the shrieking of the sacred monkey of the Hindus, and the caterwaulings of the comical little *Macacus cynemolgus.*

Almost at once Roux and Metchnikoff made an important find; their experiments were ingenious and they had about them a certain tautness and clearness that was strangely un-Metchnikoffian. Their laboratory began to be the haunt of unfortunate men who had just got syphilis; from one of these they inoculated an ape—and the very first experiment was a success. The chimpanzee developed the disease. From then on, for more than four years they toiled, transmitting the diseases from one ape to another, looking for the sneaking slender microbe but not finding it, trying to find ways to weaken the virus—as Pasteur had done with the unknown germ of rabies—in order to discover a preventive vaccine. Their monkeys died miserably of pneumonia and consumption, they got loose and ran away. While Metchnikoff, not too deftly, scratched the horrible virus into them, the apes bit him and scratched him back—and then Metchnikoff did a strange and clever experiment. He scratched a little syphilitic virus into the ear of an ape, and twenty-four hours later he cut off that ear! The ape never showed one sign of the disease in any other part of his body. . . .

"That means," cried Metchnikoff, "that the germ lingers for hours at the spot where it gets into the body—now, as in men we know exactly where the virus gets in, maybe we can kill it before it ever spreads—since in this disease we know just when it gets in, too!"

So Metchnikoff, with Roux always being careful and insisting upon good check experiments—so Metchnikoff, after all of his theorizing about why we are immune, performed one of the most profoundly practical of all the experiments of microbe hunting. He sat himself down and invented the famous calomel ointment—that now is chasing syphilis out of armies and navies the world over. He took two apes, inoculated them with the syphilitic virus fresh from a man, and then, one hour later, he rubbed the grayish ointment into that scratched spot on one of his apes. He watched the horrid signs of the disease appear on the unanointed beast, and saw all signs of the disease stay away from the one that had got the calomel.

Then for the last time Metchnikoff's strange insanity got hold of him. He forgot his vows and induced a young medical student, Maisonneuve, to volunteer to be scratched with syphilis from an infected man. Before a committee of the most distinguished medical men of France, this brave Maisonneuve stood up, and into six long scratches he watched the dangerous virus go. It was a more severe inoculation than any man would ever get in nature. The results of it might make him a thing for loathing, might send him, insane, to his death. . . . For one hour Maisonneuve waited, then Metchnikoff, full of confidence, rubbed the calomel ointment into the wounds—but not into those which had been made at the time on a chimpanzee and a monkey. It was a superb success, for Maisonneuve showed never a sign of the ugly ulcer, while the simians, thirty days afterwards, developed the disease—there was no doubt about it.

Moralists—and there were many doctors among these, mind you—raised a great clamor against these experiments of Metchnikoff. "It will remove the penalty of immorality!" said they, "to spread abroad such an easy and a perfect means of prevention!" But Metchnikoff only answered: "It has been objected that the attempt to prevent the spread of this disease is immoral. But since all means of moral prophylaxis have not prevented the great spread of syphilis and the contamination of innocents, the immoral thing is to restrain any available means we have of combating this plague."

8

Meanwhile he was scheming and groping about and having dreams about other things that might cause the arteries to harden, and suddenly he invented another cause—surely no one can say he discovered it!—"auto-intoxication, poisoning from the wild, putrefying bacilli in our large intestines—that is surely a cause of the hardening of the arteries, that is what helps us to grow old too soon!" he cried. He devised chemical tests—what awful ones they were—that would show whether the body was being poisoned from the intestine. "We would live much longer," he said, "if we had no large intestine, indeed, two people are on record, who had their large intestine cut out, and live perfectly well without it." Strange to say, he did not advocate cutting the bowels out of every one, but he set about thinking up ways of making things there uncomfortable for the "wild bacilli."

His theory was a strange one, and caused laughter and jeers and he began to get into trouble again. People wrote in, reminding him that elephants had enormous large intestines but lived to be a hundred in spite of them; that the human race, in spite of its large intestine, was one of the longest-lived species on earth. He engaged in vast obscene arguments about why evolution has allowed animals to keep a large intestine—then suddenly he hit on his great remedy for auto-intoxication. There were villages in Bulgaria where people were alleged to live to be more than a hundred. Metchnikoff didn't go down there to see—he believed it. These ancient people lived principally upon sour milk, so went the story. "Ah! there's the explanation," he muttered. He put the youngsters in his laboratory to studying the microbe that made milk sour—and in a little while the notorious Bulgarian bacillus made its bow in the rank of patent medicines.

"This germ," explained Metchnikoff, "by making the acid of sour milk, will chase the wild poisonous bacilli out of the intestine." He began drinking huge draughts of sour milk himself, and later, for years, he fed himself cultivations of the Bulgarian bacillus. He wrote large books about his new theory and a serious English journal acclaimed them to be the most important scientific treatises since Darwin's "Origin of Species." The Bulgarian bacillus became a rage, companies were formed, and their directors grew rich off selling these silly bacilli. Metchnikoff let them use his name (though Olga insists he never made a franc from that) for the label.

For nearly twenty years Metchnikoff austerely lived to the letter of his new theory. He neither drank alcoholic drinks nor did he smoke. He permitted himself no debaucheries. He was examined incessantly by the most renowned specialists of the age. His rolls were sent to him in separate sterilized paper bags so that they would be free from the wild, auto-intoxicating bacilli. He constantly tested his various juices and excretions. In those years he got down untold gallons of sour milk and swallowed billions of the beneficent bacilli of Bulgaria. . . .

And he died at the age of seventy-one.

8

Theobald Smith

Ticks and Texas Fever

1

It was Theobald Smith who made mankind turn a corner. He was the first, and remains the captain of American microbe hunters. He poked his nose—following the reasoning of some plain farmers—around a sharp turn and came upon amazing things; and now this history tells what Smith saw and what the trail-breakers who came after him found.

"It is in the power of man to make parasitic maladies disappear from the face of the globe!" So promised Pasteur, palsied but famous after his fight with the sicknesses of silkworms. He promised that, you remember, with a kind of enthusiastic vehemence, making folks think they might be rid of plagues by a year after next at the latest. Men began to hope and wait. . . . They cheered as Pasteur invented vaccines—marvelous these were but not what you would call microbe-exterminators. Then Koch came, to astound men by his perilous science of finding the tubercle bacillus, and, though Koch promised little, men remembered Pasteur's prophecy and waited for consumption to vanish. . . . Years went by while Roux and Behring battled bloodily to scotch the poison of diphtheria; mothers crooned hopeful songs into the ears of their children. . . . Some men giggled, but secretly hoped a little too, that the mighty (albeit windy) Metchnikoff might teach his phagocytes to eat up every germ in the world. . . . Diseases were getting a bit milder maybe—the reason is still mysterious—but they seemed in no hurry to vanish, and men had to keep on waiting. . . .

Then arose a young man, Theobald Smith, at the opening of the last ten years of the eighteen hundreds, to show why northern cows get sick and die of Texas fever when they go south, and to explain why southern cows, though healthy, go north and trail along with them a mysterious death for northern cattle. In 1893 Theobald Smith wrote his straight, clear report of the answer to this riddle; there was certainly no public horn-tooting about it and the report is now out of print— but that report gave an idea to the swashbuckling David Bruce; it gave hints to Patrick Manson; it set thoughts flickering through the head of the brilliant but indignant Italian, Grassi; that report gave confidence in his dangerous quest to the American Walter Reed and that gang of officers and gallant privates who refused extra pay for the job of being martyrs to research.

What kind of man is this Theobald Smith (safe to say all but a few thousand Americans have never even heard of him), and how could his discoveries about a cow disease set such dreams stirring—how could those farmer's reasonings that he proved, show microbe hunters a way to begin to realize the poetic promise of Pasteur to men?

2

In 1884 Theobald Smith was in his middle twenties; he was a Bachelor of Philosophy of Cornell University; he was a doctor of medicine from the Albany Medical College. But he detested the idea of going through life solemnly diagnosing sicknesses he could not hope to cure, offering sympathy where help was needed, trying to heal patients for whom there was no hope

—in brief, medicine seemed to him to be a mixed-up, illogical business. He was all for biting into the unknown in places where there was a chance of swallowing it—a little of it—without having mental indigestion. In short, though a physician, he wanted to do science! In especial he was eager—as what searcher was not in those piping days—about microbes. At Cornell (it was before the days of jazz) he had played psalms and Beethoven on the pipe organ; here too (college activities had not yet engulfed mere learning) Theobald Smith dug thoroughly into mathematics, into physical science, into German, and particularly he became enthusiastic about looking through microscopes. Maybe then he saw his first microbe. . . .

But when he came to the medical school at Albany, he found no excitement about possibly dastardly bacilli among the doctors of the faculty; germs had not yet been set up as targets for the healing shots of the medical profession; there was no course in bacteriology there—nor, for that matter, in any medical school in America. But he wanted to do science! And, caring nothing for the healthy drunkennesses and scientific obscenities of the ordinary medical student, Theobald Smith soothed himself with the microscopic study of the interiors of cats. In his first published paper he made certain shrewd observations on peculiar twists of anatomy in the depths of the bellies of cats—that was his bow as a searcher.

He graduated and wanted above everything to be an experimenter, but he had, before anything, to make a living. Just then young American doctors were hurrying to Europe, eager to look over Koch's shoulder to learn ways to paint bacilli, to breed them true, to shoot them under the skins of animals, and to talk like real experts about them. Theobald Smith would have liked to go but he had to find a job. And presently, while those other well-off young Americans were getting in on the ground floor of the new exciting science (afterward they told how they had actually worked in the same room with those great Germans!) and when they were getting ready to land important professorships, Theobald Smith got his job. A humble and surely not academically respectable job it was too! For he was appointed one of the staff of the then feeble, struggling, insignificant, financially rather ill-nourished, and in general almost negligible Bureau of Animal Industry at Washington. Counting Smith, there were four members of the staff of this Bureau. The Chief was a good man named Salmon. He was enthusiastically interested in what germs might do to cows and sincerely passionate about the importance of bacilli to pigs—but he knew nothing of how to find the microbes harassing these valuable creatures. Then there was Mr. Kilborne who rejoiced in the degree of Bachelor of Agriculture and was something of a horse doctor (he now runs a hardware store in New York, up-state). And finally, this staff to which Smith came, was glorified by the ancient and redoubtable Alexander, a darky ex-slave who sat about solemnly, and when urged, got up to wash the dirty bottles or chaperon the guinea-pigs.

In a little room lighted by a dormer window under the roof in the attic of a government building, Smith set out to hunt microbes. It was his proper business! Naturally he went at it, as if he had been born with a syringe in his hand and a platinum wire in his mouth. Though a university graduate, he read German well, and of nights, with gulps, he gobbled up the brave doings of Robert Koch; like a young duck taking to the water he began to imitate Koch's subtle ways of nursing and waylaying hideous bacilli and those strange spirilla who swim about like living corkscrews. . . . "I owe everything to Robert Koch!" he said, and thought of that far-off genius as some country baseball slugger might think of Babe Ruth.

In his dingy attic he was tireless. It made no difference that he was not strong—all day and part of the night he hunted microbes. And he had musician's fingers that helped him to brew microbe soups with very few spillings. In off moments he would swat the regiments of cockroaches who marched without stopping into his attic from the lumber room close by. In a remarkably short time he had taught himself everything needful and began to make cautious discoveries—he invented a queer new safe kind of vaccine, which contained no bacilli but only their filtered formless protein stuff. The heat of his attic was an intensification of the shimmering hell Washington knows how to be, but he wiped the sweat from the end of his nose and set to work in the right, classic way of Koch—with an astounding instinct he avoided the cruder methods of Pasteur.

3

You talk about freedom of science! You think a free choice to dig in any part of the Unknown is needed by searchers? I used to think so, and I have got into trouble with eminent authorities for saying so—too loudly. Wrong! For Theobald Smith, with little more freedom to start with than some low government clerk—had to research into things Dr. Salmon told him to research at, and Dr. Salmon was paid to direct Smith to solve puzzles which were bothering the farmers and stock-raisers. Such was science in the Bureau of Animal Industry. Dr. Salmon and Bachelor Kilborne and Theobald Smith—to say nothing of the indispensable Alexander—were expected to rush out like firemen and squirt science on the flaming epidemics threatening the pigs and heifers and bulls and rams of the farmers of the land. Just then the stock-raisers were seriously upset by a very weird disease, the Texas fever.

Southern cattlemen bought northern cattle; they were unloaded from their box-cars and put to graze on the fields along with perfectly healthy southern cows; everything would go well for a month or so, and then, bang! an epidemic burst out among northern cows. They stopped eating, they lost dozens of pounds a day, their urine ran strangely red, they stood aimless with arched backs and sad eyes—and in a few days every last one of the fine northern herd lay stiff-legged on the field. The same thing happened when southern steers and heifers were shipped North; they were put into northern fields, grazed there awhile, were driven away perhaps; when northern cows were turned into those fields where their southern sisters had been, in thirty days or so they began to die—in ten days after that a whole fine herd might be under the ground.

What was this strange death, brought from the South by cattle never sick with it themselves, and left invisibly in ambush on the fields? Why did it take more than a month for those fields to become dangerous? Why were they only dangerous in the hot summer months?

The whole country was excited about it; there was bad feeling between the meridional cowmen and their colleagues of the North; New York City went into a panic when carloads of stock shipped East for beef began to die in hundreds on the trains. Something must be done! And the distinguished doctors of the Metropolitan Health Board went to work to try to find the microbe cause of the disease. . . .

Meanwhile certain wise old Western cattle growers had a theory—it was just what you would call a plain hunch got from smoking their pipes over disastrous losses of cows—they had a notion that Texas fever was caused by an insect living on the cattle and sucking blood; this bug they called a *tick*.

The learned doctors of the Metropolitan Board and all of the distinguished horse doctors of the various state Experiment Stations laughed. Ticks cause disease! Any insect cause disease! It was unheard of. It was against all science. It was silly! ". . . A little thought should have satisfied any one of the absurdity of this idea," announced the noted authority, Gamgee. This man was up to his nose in the study of Texas fever, and never mentioned a tick; the scientists all over gravely cut up the carcasses of cows and discovered bacilli there (but never saw a tick). "It is the dung spreads it!" said one. "You are wrong, it is the saliva!" said another. There were as many theories as there were scientists. And the cattle kept on dying.

4

Then, in 1888, Dr. Salmon put Theobald Smith, with Kilborne to help him, and Alexander to clean up after them— saying nothing about ticks Salmon put his entire staff to work on Texas fever. "Discover the germ!" he told Smith. That year they had nothing but the spleens and livers of four dead Texas fever cows to investigate; packed in pails of ice, from Virginia and Maryland to his furnace-like attic came those livers and spleens. Theobald Smith had what so many of those mystified scientists and baffled horse doctors lacked—horse sense. He turned his microscope on to different bits of the first sample of spleen; he spied microbes in it; there was a veritable menagerie of different species of them.

Then Smith sniffed at that bit of spleen. He wrinkled up his nose—it smelled. It was spoiled.

At once he sent out messages, asking the stockmen to get the insides out of their cattle right away after they died, to pack them quickly in ice, to see they got to the laboratory more quickly. It was done, and in the next spleen he found no microbes at all—but only a great quantity of mysteriously broken up red corpuscles of the blood. "They look wrecked!" he said. But he could find no microbes. He was still young, and sarcastic, and impatient with any searcher who couldn't do close hard thinking. A man named Billings had claimed a foolish common bacillus (which he found in every part of every dead cow and in every corner of the barnyard—including the manure pile—as well) was the cause of Texas fever. Billings wrote a spread-eagle paper, saying: "The sun of original research, in disease, seems to be rising in the West instead of the East!"

"Somewhat pompous claims," said Smith, and he blew away all that pseudo-scientific rubbish in a few dry sentences. Smith knew it was no good sitting in a laboratory, with no matter how many guinea-pigs and what an array of fine syringes, simply to peer at the spleens and livers of more or less odoriferous cows. He was an experimenter; he must study the living disease; be there while the cows kicked their last quivering spasms; he must follow nature. He began to get ready for the summer of 1889, when, one day, Kilborne told him of the cattlemen's ridiculous theory about the ticks.

In a moment he pricked up his mental ears. "The farmers,

the ones who lose the stock, who see most of Texas fever, they think that?"

Now, though Theobald Smith was born in a city, he liked the smell of hay just cut and the brown furrows of fresh-turned fields. There was something sage—something as near as you can come to *truth* for him in a farmer's clipped sentences about the crops or the weather. Smith was learned in the marvelous shorthand of mathematics; men of the soil don't know that stuff. He was absolutely at home among the scopes and tubes and charts of shining laboratories—in short, this young searcher was full of sophisticated wisdom that laughs at common sayings, that often jeers at peasant platitudes. But in spite of all his learning (and this was an arbitrary strange thing about him!) Theobald Smith did not confuse fine buildings and complicated apparatus with clear thinking—he seemed always to be distrusting what he got out of books or what he saw in tubes. . . . He felt the dumbest yokel to be profoundly right when that fellow took his corn-cob pipe from his maybe unbrushed teeth to growl that April showers brought May flowers.

He listened to Kilborne's gossip about that idiotic theory of ticks; Kilborne told him the cattlemen of the West were pretty well agreed it was ticks. Well, pondered Smith, those fellows were surely innocent of any fancy reasoning to corrupt their brains, they reeked of the smell of steers and heifers, they were almost, you might say, a part of their animals; and they were the ones who had to lay awake nights knowing this dreadful disease was turning their cattle's blood to water, to taking the bread from their children's mouths. They had to bury those poor wasted beasts. And these experienced farmers one and all said: "No ticks—no Texas fever!"

Theobald Smith would follow the farmers. He would watch the disease as nearly as possible as those stockmen had watched it. Here was a new kind of microbe hunting—following nature, and changing her by just the smallest tricks. . . . The summer of 1889 came, the days grew hot; the year before the cattlemen had complained bitterly about their losses. It was urgent to do something, even the government saw that. The Department of Agriculture loosened up with a good appropriation, and Dr. Salmon, the Director, directed that the work begin—luckily he knew so little about experiments that his direction never bothered Smith in the slightest.

5

With Kilborne, Theobald Smith now built an outlandish laboratory, not between four walls but under the hot sky, and the rooms of that place of science were nothing more than five or six little dusty fenced off fields. On June 27 of 1889, seven rather thin but perfectly healthy cows came off a little boat which brought them from farms in North Carolina, from the heart of the Texas fever country, where it was death for northern cattle to go. And these seven cows were, one and all of them, decorated, infested and plagued by several thousands of ticks, assorted sizes of them, some so tiny they needed a magnifying glass to be seen—and then there were splendid female ticks half an inch long, puffed up with blood sucked from their long-suffering hosts.

Into securely fenced Field No. 1, Smith and Kilborne drove four of these tick-loaded southern cattle, and with them they put six healthy northern beasts— "Pretty soon the northerners will be getting the ticks on them too, they have never been near Texas fever. . . . They are susceptible, and then . . . ?" said Smith. "And now for a little trick to see if it is the ticks we have to blame!"

So Theobald Smith did his first little trick—call it an experiment if you wish—it was a stunt a shrewd cattleman might have thought of if he hadn't been too busy to try it; it was an experiment all other American scientists considered it silly to attempt. Smith and Kilborne set out to pick off, with their hands, every single tick from the remaining three southern cattle! The beasts kicked and switched their tails in these strange experimenters' faces; it was way over a hundred in the sun, and the dust from the rampaging of the offended cows hung in clouds around them and stuck to their sweaty fore-heads. Buried away under the matted hair of the cattle hid those ticks, and the little ones out in the open seemed to crawl away under the hair when the cramped fingers of the searchers went after them. And how those damned parasites stuck to their cow-hosts—there were magnificent blood-gorged lady ticks who mashed up into nasty messes when you tried to pull them off—it was a miserable business!

But toward evening of that day they could find never a tick on any of those three North Carolina cows, and into Field No. 2 they put them, along with four healthy northern beasts. "These northerners, perfectly fit for a fatal attack of Texas fever, will be rubbing noses with the southerners, will be nibbling the same grass, drinking from the same water, sniffing at the North Carolina cow's excretions—but they'll get no ticks from them. Well—now to wait and see if it's the ticks who are to blame!"

July and the first of August were two months of hot but strenuous waiting. Smith, with a Government bug-expert named Cooper Curtice, kept himself busy with vast studies of the lives and works and ways of ticks. They discovered how a six-legged baby tick climbs up onto a cow, how it fastens itself to the cow's hide, begins to suck blood, sheds its skin, proudly acquires two more legs, sheds its skin again; they found out the eight-legged females then marry (on the cow's back) each of them a little male, how the lady-ticks then have great feasts of blood, grow to tick womanhood—and at last drop off the cow to the ground to lay their two thousand or more eggs; so, hardly more than twenty days after their journey up the leg of the cow, their mission in life is done, and they shrivel up and die—while strange doings begin in each of those two thousand eggs. . . .

Meanwhile, every day—it was a relief to get out of that cockroachy attic even to those burning fields—Theobald Smith journeyed out to his open air laboratory where Kilborne the future hardware dealer was in command. He went to Field No. 1 to see if ticks had got on to any of the northern cattle yet, to see if they were getting hot, if their heads drooped; he crossed over to Field No. 2 to pick a few more ticks off those three North Carolina cows—a few new ones always seemed to be popping up, grown from ones too small to see that first day!—it was nervous business, making sure those three cows stayed clean of ticks. . . . It was, to tell the truth, a perspiring and not too interesting waiting until that day a little past the middle of August, when the first northern cow began to show ticks, and presently to stand with her back arched, refusing to eat. Then the ticks appeared on all the northerners; they burned with fever, their blood turned to water, their ribs stuck out and their flanks grew bony—and ticks? They seemed to be alive with ticks!

But on Field No. 2, where there were no ticks, the northern cows stayed as healthy as their North Carolina mates. . . .

Each day the fever of the northern beasts in Field No. 1 went higher—then one by one they died; the barns ran red with the blood of the post mortems, and there were rushings to and fro between the dead beasts on the field and the microscopes in the attic—even Alexander, dimly sensing the momentous things afoot, even Alexander got busy. And Theobald Smith looked at the thin blood of the dead cows. "It is the blood the unknown Texas fever microbe attacks—something seems to get into the blood corpuscles of the cows and burst them open—it is *inside* the blood cells I must look for the germ," pondered Smith. Now, though he distrusted the reports of alleged microscope experts, he was nevertheless himself mighty sharp with this machine. He turned his most powerful lens onto the blood of the first cow that died, and—here was luck!—in the very first specimen he spied queer little punched-out pear-shaped spaces in the otherwise solid discs of the blood corpuscles. At first they simply looked like h les, but he focussed up and down, and fussed, and looked at a dozen thin bits of glass with blood between them. Presently these spaces began to turn into queer pear-shaped living creatures for him. In the blood of every beast dead of Texas fever he found them—always inside the corpuscles, wrecking the corpuscles, turning the blood to water. Never did he find them in the blood of a healthy northern cow. . . . "It may be the microbe of Texas fever," he whispered, but like a good peasant he did not jump to conclusions—he must look at the blood of a hundred cows, sick and healthy, he must examine millions of red blood cells to be sure. . . .

By now the hottest weather had passed, it was September, and in Field No. 2, the northern cattle, all four of them, kept on grazing and grew fat—there were no ticks there. And Smith muttered: "We'll see if it's the ticks who are to blame!" and he took two of these unharmed northern beasts and led them into Field No. 1, where so many beasts had died—in a week a few of the little red-brown bugs were crawling up these new cows' legs. In a little more than two weeks one of these cows was dead, and the other sick, of Texas fever.

But there never was a man who needed more experiences to convince him of something he wanted to believe. He must be sure! And there was still another simple trick he could try —call it an experiment if you wish. From North Carolina, from the fatal fields down there, came large cans and these cans were filled with grass, that swarmed with ticks, crawling, thirsty for the blood of cows. These cans Theobald Smith took on to Field No. 3, where no southern cattle or their blood-sucking parasites had ever been, and he plodded up and down this field, and all over it he sowed his maybe fatal seed—of ticks. Then four northern cattle were led by Kilborne on to this field—and in a few weeks their blood ran thin, and one died, and two of the remaining three had severe bouts of Texas fever but recovered.

6

So, first of all microbe hunters, Theobald Smith traced out the exact path by which a sub-visible assassin goes from one animal to another. In the field where there were southern cattle and ticks, the northern cattle died of Texas fever; in the field where there were southern cattle *without* ticks the northern cows grew fat and remained, happy; in the field where there were no southern cattle but *only* ticks—there too, the northern cattle came down with Texas fever. It must be the tick. By such simple, two-plus-two-make-four—but oh! what endlessly careful experiments, Theobald Smith proved those western cowmen to have observed a great new fact of nature.... He chiseled that fact out of folk-shrewdness, just as the anonymous invention of the wheel has been taken out of folk-inventiveness and put to the uses of modern whirring dynamos....

You would think he thought he had proved enough—those experiments were so clear. You would think he would have advised the government to start an exterminating war on ticks, but that was not the kind of searcher Theobald Smith was. Instead, he waited for the heat of the summer of 1890 to come, and then he started doing the same experiments over, and some new ones too, all of them simple tricks, but each of them necessary to nail down the fact that the tick was the real criminal. "How do those bugs carry the disease from a southern cow to a northern one?" he pondered. "We know now one tick lives its whole life on just one cow—it doesn't flit from beast to beast like a fly...." This was a knotty question—too subtle for the crude science of the ranchers—and Smith set himself to chew that knot....

"It must be," he meditated, "that ticks, when they have sucked enough blood, and are ripe, drop off, and are crushed, and leave the little pear-shaped microbes on the grass—to be eaten by the northern cattle!"

So he took thousands of ticks, sent up in those cans from North Carolina, and mixed them with hay, and fed them to a susceptible northern cow kept carefully in a special stable. But nothing happened; the cow seemed to relish her new food; she got fat. He tried drenching another cow with mashed up ticks made into a soup—but that cow too seemed to enjoy her strange dose. She prospered on it.

It was no go—cows didn't, apparently, get the microbe by

eating ticks; he was mixed up for a while. And other plaguey questions kept him awake nights. Why was it that it took thirty days or more, after the southern tick-loaded cows came on the field, for such a field to become dangerous? Stockmen knew this too; they knew they could mix just-arrived southern cows with northern ones, and keep them together twenty days or so, and then if they took the northern ones away—they would never get Texas fever; but if you left them in that field a little longer (even if the southern cows were taken away) bang! would come the fatal epidemic into the herd of northerners. That was a poser!

Then one day in this summer of 1890, by the most strange, the most completely unforeseen of accidents, every jagged piece of the puzzle fell into its proper place. The solution of the riddle fairly clubbed Theobald Smith; it yelled at him; it forced itself on him while he was busy doing other things. He was at all kinds of experiments just then; he was bleeding northern cows for gallons of blood to give them an anemia— to make sure those funny little pear-shaped objects he had found in the corpuscles of Texas fever cattle were microbes, and not simply little changes in blood that might come from anemia. He was learning to hatch nice clean young ticks ar- tificially in glass dishes in his laboratory; he was still labori- ously picking ticks off southern cows—and sometimes he failed to get them all off and the experiments went wrong— to prove that tickless southern cows are harmless to northern ones; he was discovering the strange fact that northern calves get only a mild fever on a field fatal to their mothers. He fussed about finding every single effect a tick might have on a northern cow—it might do other damages besides giving her Texas fever . . . ?

Then came that happy accident. He asked himself: "If I should put good clean young ticks, hatched in glass dishes in my attic, ticks who never have been on cattle or on a danger- ous field—if I should put such ticks on a northern cow and let them suck their fill of her blood—could those ticks take out enough blood to give the cow an anemia?" It seems to me to have been an aimless question. His thoughts were a thousand miles away from Texas fever. . . .

But he tried it. He took a good fat yearling heifer, put her in a box-stall, and day after day put hundreds of clean baby ticks on her, holding her while these varmints crawled away beneath her hair to get a good grip on her hide. Then day after day, while the ticks made their meals, he cut little gashes in her skin to get a drop of blood to see if she was becoming anemic. And one morning Theobald Smith came into her stall—for the usual routine—he put his hand on that heifer. . . . What was this? She felt hot! Very hot! Suspiciously too hot! She dropped her head, and would not eat—and her blood which before had welled out from the gashes thick and rich and red—that blood ran very thin and darkish. He hurried back to his attic with samples of the blood between little pieces of glass. . . . Under the microscope it went, and sure enough! —here were twisted, jagged, wrecked blood corpuscles instead of good even round ones with edges smooth as a worn dime. And inside these broken cells—it was fantastical, this business!—were the little pear-shaped microbes. . . . Here was the fact, stranger than any pipe-dream—for these microbes must have come up from North Carolina on old ticks, had gone out of the old ticks into the eggs they had laid in the glass dishes, they had survived in the baby ticks hatched out of these eggs—and these babies had at last shot them back, ready to kill, into their destined but completely accidental victim, that yearling heifer!

In a flash all those mysterious questions cleared up for Theobald Smith.

It was not the old, blood-stuffed tick but its child, the baby tick, who sneaked the assassin into the northern cows; it was this little five- or ten-day-old bug who carried the murderer.

Now he saw why it was that fields took so long to become dangerous—the mother ticks have to drop off the southern cattle; it takes them some days to lay their eggs; these eggs take twenty days or more to hatch; the tick babies have to scamper about to find a cow's leg to crawl up on—all that takes many days, weeks. Never was there a simpler answer to a problem which, without this strange chance, might not yet be solved. . . .

So soon as he could hatch out other thousands of ticks in warm glass dishes, Theobald Smith proceeded to confirm his marvelous discovery; he proved it clean. For every northern cow, on whom he stuck his regiments of incubator ticks, came down with Texas fever. But he was a glutton for proofs, as you have seen, and when the summer of 1890 waned and it grew cold, he installed a coal-stove in a stable, hatched the ticks in a heated place, put a cow in the hot stable, stuck the little ticks diligently onto the hide of the cow, the stove instead of the sun made them grow as they should—and the cow got Texas fever in the winter, a thing which never happens in nature!

For two more summers Smith and Kilborne tramped about their fields, caulking up every seam in the ship of their research, answering every argument, devising astounding simple but admirably adequate answers to every objection the savant horse doctors might make—before these critics ever had a chance to make objections. They found strange facts about immunity. They saw northern calves get mild attacks of Texas fever, a couple of attacks in one summer maybe, and then next year, more or less grown up, graze unconcerned on fields absolutely murderous to a non-immune northern cow. . . . So they explained why southern cattle never die of Texas fever. This fell disease is everywhere that ticks are in the South— and ticks are everywhere; ticks are biting southern cattle and shooting the fatal queer pears into them all the time; these cattle carry the microbes about with them in their blood—but it doesn't matter, for the little sickness in their calfhood has made them immune.

Finally, after four of these stifling but triumphant summers, Theobald Smith sat down, in 1893, to answer all the perplexing questions about Texas fever—and to tell how the disease can be absolutely wiped out (just then the ancient Pasteur who had prophesied that about *all* disease was getting ready to die).

Never—and I do not forget the masterpieces of Leeuwenhoek or Koch or any genius in the line of microbe hunters—never, I say, has there been written a more simple but at the same time more solid answer to an enigma of nature. A bright boy could understand it; Isaac Newton would have taken off his hat to it. He loved Beethoven, did young Smith, and for me this "Investigation into the Nature, Causation, and Prevention of Texas or Southern Cattle Fever" has the quality of that Eighth Symphony of Beethoven's sour later years. Absurdly simple in their themes they both are, but unearthly varied and complete in the working out of those themes—just as nature is at once simple and infinitely complex. . . .

7

And so, with this report, Theobald Smith made mankind turn a corner, showed men an entirely new and fantastic way a disease may be carried—by an insect. And only by that insect. Wipe out that insect, dip all of your cattle to kill all their ticks, keep our northern cattle in fields where there are no ticks, and Texas fever will disappear from the earth. To-day whole states are dipping their cattle and to-day Texas fever which once threatened the great myriads of American cattle is no longer a matter for concern. But that is only the beginning of the beneficent deeds of this plain report, this classic unappreciated and completely out of print. For presently, on the veldt and in the dangerous bush of southern Africa, a burly Scotch surgeon-major swore at the bite of a tsetse fly—and wondered what else besides merely annoying one, these tsetse flies might do. And a little later in India, and at the same time in Italy, an Englishman and an Italian listened to the whining song of swarms of mosquitoes, and dreamed and wondered and planned strange experiments—

But those are the stories the next chapters will celebrate. They tell of ancient plagues now in reach of mankind's complete control—they tell of a deadly yellow disease now almost entirely abolished. They tell of men projecting pictures of swarming human life and turreted cities of the future reaching up and up, built on jungles now fit only for man-killing wild beasts and lizards. It was this now nearly forgotten microbe hunting of Theobald Smith that first gave men the right to have visions of a world transformed.

9

Bruce

Trail of the Tsetse

1

"Young man!"—the face of the Director-General of the British Army Medical Service changed from an irritated red to an indignant mauve-color—"young man, I will send you to India, I will send you to Zanzibar, I will send you to Timbuctoo— I will send you anywhere I please"—(the majestic old gentleman was shouting now, and his face was a positively furious purple) "but you may be damned sure I shall not send you to Natal! . . ." Reverberations. . . .

What could David Bruce do, but salute, and withdraw from his Presence? He had schemed, he had begged, and pulled wires, finally he had dared the anger of this Jupiter, so that he might go hunt microbes in South Africa. It was in the early eighteen nineties; Theobald Smith, in America, had just made that revolutionary jump ahead in microbe hunting—he had just shown how death may be carried by a tick, and only by a tick, from one animal to another. And now this David Bruce, physically as adventurous as Theobald Smith was mildly professorial, wanted to turn that corner after Smith. . . . Africa swarmed with mysterious viruses that made the continent a hell to live in; in the olive-green mimosa thickets and the jungle hummed and sizzled a hundred kinds of flies and ticks and gnats. . . . What a place for discoveries, for swashbuckling microscopings and lone-wolf bug-huntings Africa must be!

It was in the nature of David Bruce to do things his superiors and elders didn't want him to do. Just out of medical school in Edinburgh, he had joined the British Army Medical Service, not to fight, nor to save lives, nor (at that time) to get a chance to hunt microbes—not for any such noble objects. He had joined it because he wanted to marry. They hadn't a shilling, neither Bruce nor his sweetheart; their folks called them thirteen kinds of romantic idiots—why couldn't they wait until David had established himself in a nice practice?

So Bruce joined the army, and married on a salary of one thousand dollars a year.

In certain ways he was not a model soldier. He was disobedient, and, what is much worse, tactless. Still a lieutenant, he one day disapproved of the conduct of his colonel, and offered to knock him down. . . . If you could see him now, past seventy, with shoulders of a longshoreman and a barrel-chest sloping down to his burly equator, if you could hear him swear through a mustache Hindenburg would be proud to own, you would understand he could, had it been necessary, have put that colonel on his back, and laughed at the court-martial that would have been sure to follow. He was ordered to the English garrison on the Island of Malta in the Mediterranean; with him went Mrs. Bruce—it was their honeymoon. Here again he showed himself to be things soldiers seldom are. He was energetic, as well as romantic. There was a mysterious disease in the island. It was called Malta fever. It was an ill that sent pains up and down the shin bones of soldiers and made them curse the day they took the Queen's shilling. Bruce saw it was silly to sit patting the heads of these sufferers, and futile to prescribe pills for them—he must find the cause of Malta fever!

So he got himself into a mess. In an abandoned shack he set up a laboratory (little enough he knew about laboratories!)

and here he spent weeks learning how to make a culture me-
dium, out of beef broth and agar-agar, to grow the unknown
germ of Malta fever in. It ought to be simple to discover it.
His ignorance made him think that; and in his inexperience
he got the sticky agar-agar over hands and face; it stained his
uniform; the stuff set into obstinate jelly when he tried to filter
it; he spent weeks doing a job a modern laboratory helper
would accomplish in a couple of hours. He said unmentionable
things; he called Mrs. Bruce from the tennis lawn, and demanded (surely any woman knew better how to cook) that she
help him. Out of his thousand dollars a year he bought
monkeys—improvidently—at one dollar and seventy-five
cents apiece. He tried to inject the blood of the tortured soldiers into these creatures; but they wriggled out of his hands
and bit him and scratched him and were in general infernally
lively nuisances. He called to his wife: "Will you hold this
monkey for me?"

That was the way she became his assistant, and as you will
see, for thirty years she remained his right hand, going with
him into the most pestilential dirty holes any microbe hunter
has ever seen, sharing his poverty, beaming on his obscure
glories; she was so important to his tremendous but not notorious conquests. . . .

They were such muddlers at first, it is hard to believe it,
but together these newly wed bacteriologists worked and discovered the microbe of Malta fever—and were ordered from
Malta for their pains. "What was Bruce up to, anyway?" So
asked the high medical officers of the garrison. "Why wasn't
he *treating* the suffering soldiers—what for was he sticking
himself away there in the hole he called his laboratory?"
And they denounced him as an idiot, a visionary, a good-for-
nothing monkey-tamer and dabbler with test-tubes. And just—
he did do this twenty years later—as he might have discovered
how the little bacillus of Malta fever sneaks from the udders
of goats into the blood of British Tommies, he was ordered
away to Egypt.

2

Then he was ordered back to England, to the Army Medical School at Netley, to teach microbe hunting there—for hadn't he discovered the germ of an important disease? Here he met (at last God was good to him) His Excellency, the Honorable Sir Walter Hely-Hutchinson, Governor of Natal and Zululand, et cetera, et cetera. Together these two adventurers saw visions and made plans. His Excellency knew nothing about microbes and had perhaps never heard of Theobald Smith—but he had a colonial administrator's dream of Africa buzzing with prosperity under the Union Jack. Bruce cared no fig for expansion of the Empire, but he knew there must be viruses sneaking from beast to beast and man to man on the stingers of bugs and flies. He wanted (and so did Mrs. Bruce) to investigate strange diseases in impossible places.

It was then that he, only a brash captain, went to the majestic Director-General, and I have just told how he was demolished. But even Directors-General cannot remember the uppish wishes of all of their pawns and puppets; directors may propose, but adroit wire-pulling sometimes disposes, and presently in 1894, Surgeon-Major David Bruce and Mrs. Bruce are in Natal, traveling by ox-team ten miles a day towards Ubombo in Zululand. The temperature in the shade of their double-tent often reached 106; swarms of tsetse flies escorted them, harassed them, flopped on them with the speed of express trains and stung them like little adders; they were howled at by hyenas and growled at by lions. . . . They spent part of every night scratching tick bites. . . . But Bruce and his wife, the two of them, were the First British Nagana Commission to Zululand. So they were happy.

They were commanded to find out everything about the disease called nagana—the pretty native name for an unknown something that made great stretches of South Africa into a desolate place, impossible to farm in, dangerous to hunt big game in, suicidal to travel in. Nagana means "depressed and low in spirits." Nagana steals into fine horses and makes their coats stare and their hair fall out; while the fat of these horses melts away nagana grows watery pouches on their bellies and causes a thin rheum to drip from their noses; a milky film spreads over their eyes and they go blind; they droop, and at last die—every last horse touched by the nagana dies. It was the same with cattle. Farmers tried to improve their herds by importing new stock; cows sent to them fat and in prime condition came miserably to their kraals—to die of nagana. Fat droves of cattle, sent away to far-off slaughter-houses, arrived there hairless, hidebound skeletons. There were strange belts of country through which it was death for animals to go. And the big game hunters! They would start into these innocent-seeming thickets with their horses and pack-mules; one by one—in certain regions mind you—their beasts wilted under them. When these hunters tried to hoof it back, sometimes they got home.

Bruce and Mrs. Bruce came at last to Ubombo—it was a settlement on a high hill, looking east toward the Indian Ocean across sixty miles of plain, and the olive-green of the mimosa thickets of this plain was slashed with the vivid green of glades of grass. On the hill they set up their laboratory; it consisted of a couple of microscopes, a few glass slides, some knives and syringes and perhaps a few dozen test-tubes—smart young medical students of to-day would stick up their noses at such a kindergarten affair! Here they set to work, with sick horses and cattle brought up from the plain below—for Providence had so arranged it that beasts could live on the barren hill of Ubombo, absolutely safe from nagana, but just let a farmer lead them down into the juicy grass of that fertile plain, and the chances were ten to one they would die of nagana before they became fat on the grass. Bruce shaved the ears of the horses and jabbed them with a scalpel, a drop of blood welled out and Mrs. Bruce, dodging their kicks, touched off the drops onto thin glass slides.

It was hot. Their sweat dimmed the lenses of their microscopes; they rejoiced in necks cramped from hours of looking;

they joked about their red-rimmed eyes. They gave strange nicknames to their sick cows and horses, they learned to talk some Zulu. It was as if there were no Directors-General or superior officers in existence, and Bruce felt himself for the first time a free searcher.

And very soon they made their first step ahead: in the blood of one of their horses, sick to death, Bruce spied a violent unwonted dancing among the faintly yellow, piled-up blood corpuscles; he slid his slide along the stage of his microscope, till he came to an open space in the jungle of blood cells. . . .

There, suddenly, popped into view the cause of the commotion—a curious little beast (much bigger than any ordinary microbe though), a creature with a blunt rear-end and a long slim lashing whip with which he seemed to explore in front of him. A creature shaped like a panatella cigar, only it was flexible, almost tying itself in knots sometimes, and it had a transparent graceful fin running the length of its body. Another of the beasts swam into the open space under the lens, and another. What extraordinary creatures! They didn't go

stupidly along like common microbes—they acted like intelligent little dragons. Each one of them darted from one round red blood cell to another; he would worry at it, try to get inside it, tug at it and pull it, push it along ahead of him—then suddenly off he would go in a straight line and bury himself under a mass of the blood cells lining the shore of the open space. . . .

"Trypanosomes—these are!" cried Bruce, and he hurried to show them to his wife. In all animals sick with nagana they found these finned beasts, in the blood they were, and in the fluid of their puffy eyelids, and in the strange yellowish jelly that replaced the fat under their skins. And never a one of them could Bruce find in healthy dogs and horses and cows. But as the sick cattle grew sicker, these vicious snakes swarmed more and more thickly in their blood, until, when the animals lay gasping, next to death, the microbes writhed in them in quivering masses, so that you would swear their blood was made up of nothing else. . . . It was horrible!

But how did these trypanosomes get from a sick beast to a healthy one? "Here on the hill we can keep healthy animals in the same stables with the sick ones—and never a one of the sound animals comes down . . . here on the hill no cow or horse has ever been known to get nagana!" muttered Bruce. "Why? . . ."

He began to dream experiments, when the long arm of the Authorities—maybe it was that dear old Director-General remembering—found him again: Surgeon-Major Bruce was to proceed to Pietermaritzburg for duty in the typhoid epidemic raging there.

3

Only five weeks they had been at this work, when they started back to Pietermaritzburg, ten miles a day by ox-team through the jungle. He started treating soldiers for typhoid fever, but as usual—thief that he was—he stole time to try to find out something about typhoid fever, in a laboratory set up, since there was no regular one, of all places—in the morgue. There in the sickening vapors of the dead-house Bruce puttered in snatched moments, got typhoid fever himself, nearly died, and before he got thoroughly better was sent out as medical officer to a filibustering expedition got up to "protect" a few thousand square miles more of territory for the Queen. It looked like the end for him, Hely-Hutchinson's wires got tangled—there seemed no chance ever to work at nagana again; when the expedition had pierced a couple of hundred miles into the jungle, all of the horses and mules of this benevolent little army up and died, and what was left of the men had to try to hoof it back. A few came out, and David Bruce was among the lustiest of those gaunt hikers. . . .

Nearly a year had been wasted. But who can blame those natural enemies of David Bruce, the High Authorities, for keeping him from research? They looked at him; they secretly trembled at his burliness and his mustaches and his air of the Berserker. This fellow was born for a soldier! But they were so busy, or forgot, and presently Hely-Hutchinson did his dirty work again, and in September, 1895, Bruce and his wife got back to Ubombo, to try to untangle the knot of how nagana gets from a sick animal to a healthy one. And here Bruce followed, for the first time, Theobald Smith around that corner. . . . Like Theobald Smith, Bruce was a man to respect and to test folk-hunches and superstitions. He respected the beliefs of folks, himself he had no fancy super-scientific thoughts and never talked big words—yes, he respected such hunches—but he must test them!

"It is the tsetse flies cause nagana," said some experienced Europeans. "Flies bite domestic animals and put some kind of poison in them."

"Nagana is caused by big game," said the wise Zulu chiefs and medicine men. "The discharges of the buffalo, the quagga, and waterbuck, the kudu—these contaminate the grass and the watering places—so it is horses and cattle are hit by the nagana."

"But why do we always fail to get our horses safe through the fly country—why is nagana called the fly disease?" asked the Europeans.

"Why, it's easy to get animals through the fly belt so long as you don't let them eat or drink!" answered the Zulus.

Bruce listened, and then proceeded to try out both ideas. He took good healthy horses, and tied heavy canvas bags round their noses so they couldn't eat nor drink; he led them down the hill to the pleasant-looking midday hell in the mimosa thickets; here he kept them for hours. While he watched to see they didn't slip their nose bags, swarms of pretty brown and gold tsetses buzzed around them—flopped on to the kicking horses and in twenty seconds swelled themselves up into bright balloons of blood. . . . The world seemed made of tsetse flies, and Bruce waved his arms. "They were enough to drive one mad!" he told me, thirty years afterward. I can see him, talking to those pests in the language of a dock-foreman, to the wonder of his Zulus. Day after day this procession of Bruce, the Zulus, and the experimental horses went down into the thorns, and each afternoon, as the sun went down behind Ubombo, Bruce and his migrating experiment grunted and sweated back up the hill.

Then, in a little more than fifteen days, to the delight of Bruce and his wife, the first of those horses who had served as a fly-restaurant turned up seedy in the morning and hung his head. And in the blood of this horse appeared the vanguard of the microscopic army of finned wee devils—that tussled so intelligently with the red blood cells. . . .

So it was with every horse taken down into the mimosa— and not one of them had eaten a blade of grass nor had one swallow of water down there; one and all they died of the nagana.

"Good, but it is not proved yet, one way or another," said Bruce. "Even if the horses didn't eat or drink, they may have *inhaled* those trypanosomes from the air—that's the way the greatest medical authorities think malaria is passed on from one man to the next—though it sounds like rot to me." But for Bruce nothing was rot until experiment proved it rot. "Here's the way to see," he cried. "Instead of taking the horses down, I'll bring the flies up!"

So he bought more healthy horses, kept them safe on the hill, thousands of feet above the dangerous plain, then once more he went down the hill—how that man loved to hunt, even for such idiotic game as flies!—and with him he took a decoy horse. The tsetses landed on the horse; Bruce and the Zulus picked them off gently, hundreds of them, and stuck them into an ingenious cage, made of muslin. Then back up the hill, to clap the cage buzzing with flies on to the back of a healthy horse. Through a clever glass window in one of the cage-sides they watched the greedy brutes make their meal by sticking their stingers through the muslin. And in less than a month it was the same with these horses, who had never eaten, nor drunk, nor even inhaled the air of the plain—every one died of the nagana.

How they worked, Bruce and his wife! They post-mortemed dead horses; they named a sick horse "The Unicorn" and tried to keep him alive with arsenic. To find out how long a tsetse fly can carry the trypanosomes on his stinger they put cages of flies on sick dogs and then at intervals of hours, and days, let them feed on healthy ones. They fed dying heifers hot pails of coffee, mercifully they shot dogs thinned by the nagana to sad bags of bones. Mrs. Bruce sterilized silk threads, to dip in blood swarming with trypanosomes, then sewed these threads under the hides of healthy dogs—to find out how long such blood might remain deadly. . . . There was now no doubt the tsetse flies, and only the flies, could carry the nagana, and now Bruce asked:

"But where do the tsetses of the plain *get* the trypanosomes they stick into cows and horses? In those fly belts there are often no horses or cattle sick with nagana, for months. Surely the flies [he was wrong here] can't stay infected for months—it must be they get them from the wild animals, the big game!" That was a possibility after his heart. Here was a chance to do something else than sit at a microscope. He forgot instantly about the more patient, subtle jobs that demanded to be done—teasing jobs, for a little man, jobs like tracing the life of the trypanosomes in the flies. . . . "The microbes must be in game!" and he buckled on his cartridge belt and loaded his guns. Into the thickets he went, and shot Burchell's zebras; he brought down kudus and slaughtered water-bucks. He slashed open the dead beasts and from their hot hearts sucked up syringes full of blood, and jogged back up the hill with them. He looked through his microscopes for trypanosomes in these bloods—but didn't find them. But there was a streak of the dreamer in him. "They may be there, too few to see," he muttered, and to prove they were there he shot great quantities of the blood from ten different animals into healthy dogs. So he discovered that the nagana microbes may lurk in game, waiting to be carried to gentler beasts by the tsetse. So it was Bruce made the first step towards the opening up of Africa.

4

And Hely-Hutchinson saw how right he had been about David Bruce. " 'Ware the tsetse fly," he told his farmers, "kill the tsetse fly, clear the thickets in which it likes to breed—drive out, exterminate the antelope from which it sucks the trypanosomes." So Bruce began ridding Africa of nagana.

Then came the Boer War. Bruce and Mrs. Bruce found themselves besieged in Ladysmith with nine thousand other Englishmen. There were thirty medical officers in the garrison—but not one surgeon. With each whine and burst of the shells from the Boer's "Long Tom" the rows of the wounded grew—there were moanings, and a horrid stench from legs that should be amputated. . . . "Think of it! Not one of those medicoes could handle a knife! Myself, I was only a laboratory man," said Bruce, "but I had cut up plenty of dogs and guinea-pigs and monkeys—so why not soldiers? There was one chap with a bashed-up knee . . . well, they chloroformed him, and while they were at that, I sat in the next room reading Treve's Surgery on how to take out a knee-joint. Then I went in and did it—we saved his leg." So Bruce became Chief Surgeon, and fought and starved, nearly to death, with the rest. What a boy that Bruce was! In 1924 in Toronto, in a hospital as he lay propped up, a battered bronchitic giant, telling me this story, his bright eye belied his skin wrinkled and the color of old parchment—and there was no doubt he was as proud of his slapdash surgery and his sulky battles with the authorities, as of any of his discoveries in microbe hunting. He chuckled through phlegm that gurgled deep in his ancient air-tubes: "Those red-tape fellows—I always had to fight their red-tape—until at last I got too str-r-rong for them!"

5

Presently, two years after Ladysmith, he became stronger than they—and they came asking him to hunt microbes. . . .

For death was abroad on the shores of Lake Victoria Nyanza, in Central Africa, on the Equator. It crept, it jumped, it kept popping up in new villages, it was in a way a very merciful death—though slow—for it was without pain, turning from a fitful fever into an unconquerable laziness strange to see in the busy natives of the lake shore; it passed, this death, from lethargy into a ridiculous sleepiness that made the mouths of the negroes fall open while they ate; it went at last from such a drowsiness into a delicious coma—no waking from this!—and into a horrible unnatural coldness that merged with the chill of the grave. Such was the African sleeping sickness. In a few years it had killed hundreds of thousands of the people of Uganda, it had sent brave missionaries to meet their God, and English colonial administrators home to their final slumber. It was turning the most generous soil on earth back into an unproductive preserve for giraffes and hyenas. The British Colonial Office was alarmed; shareholders began to fear for their dividends; natives—those who were left—began to leave their villages of shaggy, high-pitched, thatch-roofed huts. And the scientists and doctors?

Well, the scientists and doctors were working at it. Up till now the wisest ones were as completely ignorant of what was this sleeping death as the blackest trader in bananas was ignorant. No one could tell how it stole from a black father to his neighbor's dusky pickaninnies. But now the Royal Society sent out a commission made up of three searchers; they sailed for Uganda and began researches with the blood and spinal fluid of unhappy black men doomed with this drowsy death.

They groped; they sweat in the tropic heat; they formed different opinions: one was pretty sure a curious long worm that he found in the black men's blood was the cause of this death; a second had no definite opinion that I know of; the third, Castellani, thought at first that the wee villain back of the sleeping death was a streptococcus—like the microbe that causes sore throats.

That was way off the truth, but Castellani had the merit of working with his hands, trying this, trying that, devising ingenious ways of looking at the juices of those darkies. And so one day—by one of those unpredictable stumbles that lie at the bottom of so many discoveries—Castellani happened on one of those nasty little old friends of David Bruce, a trypanosome. From inside the backbone of a deadly drowsy black man Castellani had got fluid—to look for streptococcus. He put that fluid into a centrifuge—that works like a cream separator—to try to whirl possible microbes down to the bottom of the tube in the hope to find streptococcus. Down the barrel of his microscope Castellani squinted at a drop of the gray stuff from the bottom of the fluid and saw—

A trypanosome, and this beast was very much the same type of wiggler David Bruce had fished out of the blood of horses dying of nagana. Castellani kept squinting, found more trypanosomes, in the spinal juices and even in the blood of a half a dozen doomed darkies. . . .

That was the beginning, for if Castellani had not seen them, told Bruce about them, they might never have been found.

Meanwhile the smolder of the sleeping death broke into a flare that threatened English power in Africa. And the Royal Society sent the veteran David Bruce down there, with the trained searcher Nabarro, with Staff-Sergeant Gibbons, who could do anything from building roads to fixing a microscope. Then of course Mrs. Bruce was along; she had the title of Assistant—but Bruce paid her fare.

They came down to Uganda, met Castellani. He told Bruce about the streptococcus—and the trypanosomes. Back to the laboratory went these two; microscopes were unpacked, set up; doomed darkies carried in. Heavy needles were jabbed into these sad people's spines. Castellani, the young Nabarro, and Mrs. Bruce bent over their microscopes to find the yes or no of the discovery of Castellani. There they sat, in this small room on the Equator, squinting down the barrels of their machines at a succession of gray nothingnesses.

A bellow from Bruce: "I've got one!" The rest crowd round, squint in turn, exclaim as they watch the writhing trypanosome poke his exploring whip about in the gray field of the lens. Then they go back to their places—to shout discovery in their turn. So it went, from breakfast till the swift dusk of evening. In every single sample of spinal fluid from each one of his more than forty sleeping-sickness patients, Bruce and his companions found those trypanosomes.

"But they may be in healthy people's spines too!" said Bruce. Bruce knew that if he found them in healthy negroes, all this excitement would be only a wild-goose chase—he must prove they were to be found only in folks with sleeping sickness. But to get fluid out of healthy people's spines? Folks dopey from the sleeping death didn't mind it so much—but to jab one of those big needles into the back of healthy wide-awake colored people, who had no wish to be martyrs to science. . . . Can you blame them? It is no picnic having such a spear stuck into your spine. Then Bruce hit on a crafty scheme. He went to the hospital, where there was a fine array of patients with all kinds of diseases—but no sleeping sickness—and then, flimflamming them into thinking the operation would do them good, this liar in the holy cause of microbe hunting jabbed his needles into the smalls of the backs of negroes with broken legs and with headaches, into youngsters who had just been circumcised, and into their brothers or sisters who were suffering from yaws, or the itch; from all of them he got spinal fluid.

And it was a great success. Not one of these folks—who had no sleeping sickness—harbored a single trypanosome in the fluid of their spines. Maybe the operation did do them some good—but no matter, they had served their purpose. The trypanosome, Castellani and Bruce now knew, was the cause of sleeping sickness!

Now—and this is rare in the dreamers who find fundamental facts in science—Bruce was a fiend for practical applications, not poetically like Pasteur, for Bruce wasn't given to such lofty soarings, nor was he practical in the dangerous manner of the strange genius I tell of in the last chapter of this story; but the moment he turned to the study of a new plague, Bruce's gray eyes would dart round, he would begin asking himself questions: What is the natural home of the virus of this disease?—How does it get from sick to healthy?—What is its fountain and origin?—Is there anything *peculiar* in the way this sleeping sickness has spread?

That was the way he went at it now. He had discovered the trypanosome that was the cause. There were a thousand pretty little researches to tempt the scholar in him, but he brushed all these aside. Old crafty hand at searching that he was, he fished round in his memories, and came to nagana, and screwed up his eyes: "Is there anything peculiar about the way sleeping sickness is *located* in this country?" He pondered.

He sniffed around. With Mrs. Bruce he explored the high-treed shores of the lake, the islands, the rivers, the jungle. Then the common-sense eye which sees things a hundred searchers might stumble over and go by—showed him the answer. It was strange—suspiciously strange—that sleeping sickness was only found in a very narrow strip of country—along the water, only along the water, on the islands, up the river—even by the Ripon Falls where Victoria Nyanza gives herself up to the making of the Nile, there were cases of it, but never inland. That must mean some insect, a blood-sucking insect, which lives only near water, must carry the disease. That was his guess, why, I cannot tell you. "Maybe it is a tsetsé fly, a special one living only near lake shores and river banks!"

So Bruce went around asking everybody about tsetse flies in Uganda. He inquired of local bug experts: no, they were sure tsetse flies could not live at an altitude above three thousand feet. He asked the native headmen, even the black Prime Minister of Uganda: sorry, we have a blood-sucking fly, called Kivu—but there are no tsetse flies in Uganda.

But there must be!

6

And there were. One day, as they walked through the Botanical Garden at Entebbe, Bruce pushing his bulky body between the rows of tropic plants ahead of his small wife—there was a glad shriek from her. . . . "Why, David! There are two tsetse —on your back!" That woman was a scientific Diana. She swooped on those two tsetses, and caught them, and gave them a practical pinch—just enough to kill them, and then showed them to her husband. They had been perched, ready to strike, within a few inches of his neck. Now they knew they were on the trail.

Hard work began in the laboratory; already Bruce had found an excellent experimental animal—the monkey, which he could put into a beautiful fatal sleep, just like that of a man, by injecting fluid from the spines of doomed negroes. But now to catch tsetse flies. They armed themselves with butterfly nets and the glass-windowed cages they had invented in Zululand. Then these inseparable searchers climbed into canoes; lusty crews of black boys shot them across the lake. Along the banks they walked—it was charming in the shade there—but listen! Yes, there was the buzz of the tsetse. . . . They tried to avoid being bitten. They were bit—and stayed awake nights wondering what would happen—they went back to the laboratory and clapped the cages on the backs of monkeys. It was a good time for them.

That is the secret of those fine discoveries Bruce made. It was because he was a hunter. Not only with his mind—but a bold everlastingly curious snouting hunter with his body too. If he had sat back and listened to those missionaries, or stayed listening to those bug experts—he would never have learned that Kivu was the Uganda name for the tsetse. He would never have found the tsetse. But he carried the fight to the enemy —and as for Mrs. Bruce, that woman was better than a third hand or two extra pairs of eyes for him.

Now they planned and did terrible experiments. Day after day they caused tsetse flies to feed on patients near to death (already too deep in sleep to be annoyed by the insects); they interrupted the flies in the midst of their meal, and put the angry, half-satisfied cages of them on the backs of monkeys. With all the tenderness of high-priced nurses watching over Park Avenue babies they saw to it that only their experimental flies, and no chance flies from outside, got a meal off those beasts. Other searchers might have rolled their thumbs waiting to see what happened to the monkeys, but not Bruce.

He proceeded to call in a strange gang of co-workers to help him in one of the most amazing tests of all microbe hunting. Bruce asked for an audience from the high-plumed gay-robed potentate, Apolo Kagwa, Prime Minister of Uganda. He told Apolo he had discovered the microbe of the sleeping death which was killing so many thousands of his people. He informed him many thousands more already had the parasite in their blood, and were doomed. "But there is a way to stop the ruin that faces your country, for I have reason to believe it is the tsetse fly—the insect you call Kivu—and *only* this insect, that carries the poisonous germ from a sick man to a healthy one—"

The magnificent Apolo broke in: "But I cannot believe that is so—Kivu has been on the Lake shore always, and my people have only begun to be taken by the sleeping sickness during the last few years—"

Bruce didn't argue. He bluffed, as follows: "If you do not believe me, give me a chance to prove it to you. Go down, Apolo Kagwa, to the Crocodile Point on the Lake shore where Kivu swarms so. Sit on the shore there with your feet in the water for five minutes. Don't keep off the flies—and I'll promise you'll be a dead man in two years!"

The bluff was perfect: "What then, is to be done, Colonel Bruce?" asked Apolo.

"Well, I must be dead sure I am right," Bruce told him. Then he showed Apolo a great map of Uganda. "If I'm right, where there is sleeping sickness—there we will find tsetse flies too. Where there are no tsetse—there should be no sleeping sickness."

So Bruce gave Apolo butterfly nets, and killing bottles, and envelopes; he gave directions about the exact way to set down all the facts, and he told how Apolo's darky minions might pinch the flies without getting stabbed themselves. "And then we will put our findings down on this map—and see if I'm right."

Apolo was nothing if not intelligent, and efficient. He said he would see what could be done. There were bows and amiable formalities. In a jiffy the black Prime Minister had called for his head chief, the Sekibobo, and all the paraphernalia, with rigid directions, went from the Sekibobo to the lesser headmen, and from them down to the canoe men—the wheels of that perfect feudal system were set going. . . .

Presently the envelopes began to pour in on Bruce and called him away from his monkey experiments. They cluttered the laboratory, they called him from his peerings into the intestines of tsetse flies where he looked for trypanosomes. Rapidly, with perfectly recorded facts—most of them set down by intelligent blacks and some by missionaries—the envelopes came in. It was a kind of scientific co-working you would have a hard time finding among white folks, even white medical men. Each envelope had a grubby assorted mess of biting flies, they had a dirty time sorting them, but every time they found a tsetse, a red-headed pin went into that spot on the map—and if a report of "sleeping sickness present" came with that fly, a black-headed pin joined it. From the impressive Sekibobo down to the lowest fly-boy, Apolo's men had done their work with an automatic perfection. At last the red and black dots on the map showed that where there were tsetses, there was the sleeping death—and where there were no tsetses—there was no single case of sleeping sickness!

The job looked finished. The unhappy monkeys bit by the flies who had sucked the blood of dying negroes—these monkeys' mouths fell open while they tried to eat their beloved bananas; they went to sleep and died. Other monkeys never bit by flies—but kept in the same cages, eating out of the same dishes—those monkeys never showed a sign of the disease.

Here were experiments as clean, as pretty as the best ones Theobald Smith had made. . . .

7

But now for action! Whatever of the dreamer and laboratory experimenter there was in him—and there was much—those creative parts of David Bruce went to sleep, or evaporated out of him; he became the surgeon of Ladysmith once more, and the rampageous shooter of lions and killer of kudus. . . . To wipe out the sleeping sickness! That seemed the most brilliantly simple job now. Not that there weren't countless thousands of blacks with trypanosomes in their blood, and all these folks must die, of course; not that there weren't buzzing billions of tsetses singing their hellish tune on the Lake shore—but here was the point: *Those flies lived only on the lake shore!* And if they had no more sleeping-sickness blood to suck, then. . . . And Apolo Kagwa was absolute Tsar of all Uganda . . . Apolo, Bruce knew, trusted him, adored him. . . .

Now to wipe sleeping sickness from the earth!

To conference with Bruce once more came Apolo and the Sekibobo and the lesser chiefs. Bruce told them the simple logic of what was to be done.

"Of course—that can be done," said Apolo. He had seen the map. He was convinced. He made a dignified wave of the hand to his chiefs, and gave a few words of explanation. So Bruce and Mrs. Bruce went back to England. Apolo gave his order, and then the pitiful population of black men and their families streamed inland out of the lake shore villages, away —not to return for years, or ever—from those dear shady places where they and the long line of their forefathers had fished and played and bargained and begot their kind; canoes, loaded with mats and earthen pots and pickaninnies set out (not to return) from the thickly peopled island—and the weird outlandish beating of the tom-toms no longer boomed across the water.

"Not one of you," commanded Apolo, "may live within

fifteen miles of the Lake shore—not one of you is to visit the Lake again. Then the sleeping death will die out, for the fly Kivu lives only by the water, and when you are gone she will no longer have a single sick one from whom to suck the fatal poison. When all of our people who are now sick have died, you may go back—and it will be safe to live by the Lake shore for always."

Without a word—it is incredible to us law-abiding folks—they obeyed their potentate.

The country around Lake Victoria Nyanza grew, in the frantic way tropical green things grow, back into the primordial jungle; crocodiles snoozed on the banks where big villages had been. Hippopotami waddled onto the shore and sniffed in the deserted huts. . . . The tribes of the lake, inland, were happy, for no more of them came down with that fatal drowsiness. So Bruce began to rid Africa of sleeping sickness.

It was a triumph—in a time of great victories in the fight of men against death. The secret of the spread of malaria—you will hear the not too savory story of it presently—had been found in India and Italy. And as for yellow fever—it seemed as if the yellow jack was to be put to sleep for good. Great Eminences of the medical profession pointed in speeches amid cheers to the deeds of medicine. . . . The British Empire rang with hosannahs for David Bruce. He was promoted Colonel. He was dubbed Knight Commander of the Bath. Lady Bruce? Well, she was proud of him and stayed his assistant, obscurely. And Bruce still paid, out of his miserable colonel's salary, her fare on those expeditions they were always making.

Africa looked safe for the black men, and open to the benevolent white men. But nature had other notions. She had cards up her sleeve. She almost never lets herself be conquered at a swoop, Napoleonically—as Bruce and Apolo (and who can blame them?) thought they had done. Nature was not going to let her vast specimen cabinet be robbed so easily of every last one of those pretty parasites, the trypanosomes of sleeping sickness. A couple of years passed, and suddenly the Kavirondo people, on the east shore of the Lake where sleeping death had never been—these folks began to go to sleep and not wake up. And there were disturbing reports of hunters coming down with sleeping sickness, even in those places that should have been safe, in the country from which all human life had been moved away. The Royal Society sent out another Commission (Bruce was busy with that affair of goat's milk giving Malta fever) and one of these new commissioners was a bright young microbe hunter, Tulloch. He went on a picnic one day to a nice part of the shore whose dark green was dotted with scarlet flowers. It must be safe there now, they thought, but a tsetse buzzed, and in less than a year Tulloch had drowsed into his last cold sleep. The Commission went home. . . .

Bruce—you would think he would be looking by this time for some swivel-chair button-pressing job—packed his kit-bag and went back to Uganda, to see what he had left out of those experiments that had looked so sure. He had gone off half-cocked, with that Napoleonic plan of moving a nation, but who can blame him? It had looked so simple, and how expect even the craftiest of the cheaters of Nature to find out, in a year, every single nook where Nature hides the living poisons to kill the presumptuous men who cheat her! Lady Bruce as usual went with him, and they found new epidemics of sleeping sickness flaring up in unwonted places. It was a miserable discouraging business.

Bruce was a modest man, who had no foolish vanity to tell him that his own theories were superior to brute facts. "My plan has been a washout," you can hear him grumbling. "Somewhere, aside from the human being, those tsetses must get the trypanosomes—maybe it's like the nagana—maybe they can live in wild beasts' blood too. . . ."

Now if Bruce had theories that were a little too simple he was just the same an exceedingly crafty experimenter; if he had a foolish faith in his experiments, he had the persistence to claw his way out of the bogs of disappointment that his simplicity and love of gorgeous deeds got him into. What a stubborn man he was! For, when you think of the menagerie of birds, beasts, fishes and reptiles Uganda is, you wonder why he didn't pack his bags and start back for England. But no. Once more the canoe man paddled Bruce and his lady across to that tangled shore, and they caught flies in places where for three years no man had been. Strange experiments they made in a heat to embarrass a salamander—one laborious complicated record in his notes tells of two thousand, eight hundred and seventy-six flies (which could never have bitten a human sleeping-sickness patient) fed on five monkeys—and two of these monkeys came down with the disease!

"The trypanosomes must be hiding in wild animals!" Bruce cries. So they go to the dangerous Crocodile Point, and catch wild pigs and African gray and purple herons; they bleed sacred ibises and glossy ones; they stab and get blood from plovers and kingfishers and cormorants—and even crocodiles! Everywhere they look for those deadly, hiding, thousandth-of-an-inch-long wigglers.

They caught tsetse flies on Crocodile Point. See the fantastic picture of them there, gravely toiling at a job fit for a hundred searchers to take ten years at. Bruce sits with his wife on the sand in the middle of a ring of bare-backed paddlers who squat round them. The tsetses buzz down onto the paddlers' backs. The fly-boys pounce on them, hand them to Bruce, who snips off their heads, waves the buzzing devils away from his own neck, determines the sex of each fly caught, dissects out its intestine—and smears the blood in them on thin glass slides. . . .

Washouts, most of these experiments; but one day, in the blood of a native cow from the Island of Kome, not hurting that cow at all, but ready to be sucked up by the tsetse for stabbing under the skin of the first man it meets, Bruce found the trypanosome of sleeping sickness. He sent out word, and presently a lot of bulls and cows were driven up the hill to Mpumu by order of Apolo Kagwa. Bruce, himself in the thick of it, directed dusty fly-bitings of these cattle—yes! there was no doubt the sleeping-sickness virus could live in them. Then there were scuffles in the hot pens with fresh-caught antelope;

they were thrown, they were tied, Bruce held dying monkeys across their flanks, and let harmless tsetses, bred in the laboratory, feed on the monkey and then on the buck. . . .

"The fly country around the Lake shore will have to be cleared of antelope, too, as well as men—before the Kivu become harmless," Bruce said at last to Apolo.

And now the sleeping death really disappeared from the shores of Lake Victoria Nyanza.

8

The ten thousand smaller microbe hunters who work at lesser jobs to-day, as well as the dozen towering ones whose adventures this book tells, all of them have to take some risk of death. But if the ten thousand smaller microbe hunters of to-day could by some chemistry be changed into death-fighters like Bruce! There was something diabolical in the risks he took, and something yet more devilish in the way he could laugh—with a dry humor—and wish other microbe hunters might have died to prove some of his own theories. But he had a right to wish death for others—

Can young tsetse flies, bred in the laboratory, inherit the sleeping-sickness trypanosome from their mothers? Surely there was a chance of it (you remember that strange business of Theobald Smith's mother-ticks bequeathing the Texas fever microbe to their children). But analogies are for philosophers and lawyers. "*Are* artificially hatched young tsetses dangerous?" asks Bruce. "No!" he can answer. "For two members of the commission" [modestly he does not say which two members] "allowed hundreds of tsetse flies, bred in the laboratory, to bite them. And the result was negative."

But no man knew what the result would be—before he tried. And the deaths from sleeping sickness (according to the best figures) are one hundred out of one hundred. . . .

How he enjoyed hearing of other men trying to kill themselves to find out! His last African foray was in 1911—he stayed until 1914. He was near sixty; his blacksmith's strength was beginning to crack from a nasty infection of his air-tubes got from I know not what drenching rains or chills of high tropic nights. But a new form of sleeping sickness—terrible stuff that killed in a few months instead of years—had just broken out in Nyassaland and Rhodesia. There was a great scientific quarrel on. Was the trypanosome causing this disease some new beast just out of the womb of Nature—or was it nothing else than Bruce's old parasite of nagana, tired of butchering only cows, dogs and horses, and now learning to kill men?

Bruce went to work at it. A German in Portuguese East Africa said: "This trypanosome is a new kind of bug!" Bruce retorted: "On the contrary, it is nothing but the nagana germ hopping from cows to men."

Then this German, his name was Taute, took the blood of an animal about to die from nagana, and shot five cubic centimeters of it—it held millions of trypanosomes—under his own skin: to prove the nagana parasite does not kill men. And he let scores of tsetse flies bite him, flies whose bellies and spit-glands were crammed with the writhing microbes—he did these things to prove his point!

Was Bruce shocked at this? Listen to him, then: "It is a matter for some scientific regret that these experiments were not successful—though we can ill spare our bold and somewhat rash colleague—for then the question would have been answered. . . . As it is, these negative experiments prove nothing. It may be that only one man in a thousand would become infected that way."

Merciless Bruce! Poor Taute! He tried conscientiously to kill himself—and Bruce says it is too bad he did not die. He made the ultimate gesture—surely the God of searchers will reward him; then Bruce (and he is right) criticizes the worth of Taute's lone desperate experiment!

Nyassaland was the last battlefield of Bruce against the sleeping sickness, and it was his most hopeless one. For here he found that the *Glossina morsitans* (that is the name of the tsetse carrier of the sickness) does not make its home only on the shores of lakes and rivers, but buzzes and bites from one end of Nyassaland to the other; there is no way of running away from it, no chance of moving nations out from under it here. . . . Bruce stuck at it, he spent years at measurements of the lengths of trypanosomes—monotonous enough this work was to have driven a subway ticket chopper mad—he was trying to find out whether the nagana and this new disease were one and the same thing. He ended by not finding out, and he finished with this regret: that it was *at present* impossible to do the experiment to clinch the matter one way or the other.

That experiment was the injection of the nagana trypanosomes, not into one, or a hundred—but a thousand human beings.

9

But there was grisly hope left in the old Viking. "*At present* it is impossible," he said, while he believed that somewhere, somewhen, men may be found, in the mass, who will be glad to die for truth. And as you will see, in a story of a band of American buck-privates in another chapter, there are beginnings of such spirit even now. But when great armies of men so offer themselves, to fight death, just as they now delight to fight each other, it will be because they are led on by captains such as David Bruce.

10

Ross vs. Grassi

Malaria

1

The last ten years of the nineteenth century were as unfortunate for ticks, bugs, and gnats as they were glorious for the microbe hunters. Theobald Smith had started them off by scotching the ticks that carried Texas fever; a little later and six thousand miles away David Bruce, stumbling though the African bush, got onto the trail of the tsetse fly, accused him, convicted him. How melancholy and lean have been the years, since then, for that murderous tick whose proper name is *Boophilus bovis*, and you may be sure that since those searchings of David Bruce, the tsetses have had to bootleg for the blood of black natives and white hunters, and missionaries. And now alas for mosquitoes! Malaria must be wiped from the earth. Malaria can be destroyed! Because, by the middle of 1899, two wrangling and not too dignified microbe hunters had proved that the mosquito—and only one particular kind of mosquito—was the criminal in the malaria mystery.

Two men solved that puzzle. The one, Ronald Ross, was a not particularly distinguished officer in the medical service of India. The other, Battista Grassi, was a very distinguished Italian authority on worms, white ants, and the doings of eels. You cannot put one before the other in the order of their merit—Ross would certainly have stopped short of solving the puzzle without Grassi. And Grassi might (though I am not so sure of that!) have muddled for years if the searchings of Ross had not given him hints. So there is no doubt they helped each other, but unhappily for the Dignity of Science, before the huzzahs of the rescued populations had died away, Battista Grassi and Ronald Ross were in each other's hair on the question of who did how much. It was deplorable. To listen to these two, you would think each would rather this noble discovery had remained buried, than have the other get a mite of credit for it. Indeed, the only consolation to be got from this scientific brawl—aside from the saving of human lives—is the knowledge that microbe hunters are men like the rest of us, and not stuffed shirts or sacred cows, as certain historians would have us believe. They sat there, Battista Grassi and Ronald Ross, indignant co-workers in a glorious job, in the midst of their triumph, with figurative torn collars and metaphorical scratched faces. Like two quarrelsome small boys they sat there.

2

For the first thirty-five years of his life Ronald Ross tried his best not to be a microbe hunter. He was born in the foothills of the Himalayas in India, and knowing his father (if you believe in eugenics) you might suspect that Ronald Ross would do topsy-turvy things with his life. Father Ross was a ferocious looking border-fighting English general with belligerent side-whiskers, who was fond of battles but preferred to paint landscapes. He shipped his son Ronald Ross back to England before he was ten, and presently, before he was twenty, Ronald was making a not too enthusiastic pass at studying medicine, failing to pass his examinations because he preferred composing music to the learning of Latin words and the cultivation of the bedside manner. This was in the eighteen-seventies, mind you, in the midst of the most spectacular antics of Pasteur, but from the autobiography of Ronald Ross, which is a strange mixture of cleverness and contradiction, of frank abuse of himself and of high enthusiasm for himself, you can only conclude that this revolution in medicine left Ronald Ross cold.

But he was, for all that, something of a chaser of moonbeams, because, finding that his symphonies didn't turn out to be anything like those of Mozart, he tried literature, in the grand manner. He neglected to write prescriptions while he nursed his natural bent for epic drama. But publishers didn't care for these masterpieces, and when Ross printed them at his own expense, the public failed to get excited about them. Father Ross became indignant at this dabbling and threatened to stop his allowance, so Ronald (he had spunk) got a job as a ship's doctor on the Anchor Line between London and New York. On this vessel he observed the emotions and frailties of human nature in the steerage, wrote poetry on the futility of life, and got up his back medical work. Finally he passed the examination for the Indian Medical Service, found the heat of India detestable, but was glad there was little medical practice to attend to, because it left him time to compose now totally forgotten epics and sagas and blood-and-thunder romances. That was the beginning of the career of Ronald Ross!

Not that there was no chance for him to hunt microbes in India. Microbes? The very air was thick with them. The water was a soup of them. All around him in Madras were the stinking tanks breeding the Asiatic cholera; he saw men die in thousands of the black plague; he heard their teeth rattle with the ague of malaria, but he had no ears or eyes or nose for all that—for now he forgot literature to become a mathematician. He shut himself up inventing complicated equations. He devised systems of the universe of a grandeur he thought equal to Newton's. He forgot about these to write another novel. He took twenty-five-mile-a-day walking trips in spite of the heat and then cursed India bitterly because it was so hot. He was ordered off to Burma and to the Island of Moulmein, and here he did remarkable surgical operations—"which cured most of the cases"—though he had never presumed to be a surgeon. He tried everything but impressed hardly anybody; years passed, and, when the Indian Medical Service failed to recognize his various abilities, Ronald Ross cried: "Why work?"

He went back to England on his first furlough in 1888, and there something happened to him, an event that is often an antidote to cynicism and a regulator of confused multitudinous ambitions. He met, he was smitten with, and presently he married Miss Rosa Bloxam. Back in India—though he wrote another novel called "Child of Ocean" and invented systems of shorthand and devised phonetic spellings for the writing of verse and was elected secretary of the Golf Club—he began to fumble at his proper work. In short he began to turn a microscope, with which he was no expert, on to the blood of malarious Hindus. The bizarre, many-formed malaria microbe had been discovered long ago in 1880 by a French army surgeon, Laveran, and Ronald Ross, who was as original as he was energetic and never did anything the way anybody else did it, tried to find this malaria germ by methods of his own.

Of course, he failed again. He bribed, begged, and wheedled drops of blood out of the fingers of hundreds of aguey East Indians. He peered. He found nothing. "Laveran is certainly wrong! There is no germ of malaria!" said Ronald Ross, and he wrote four papers trying to prove that malaria was due to intestinal disturbances. That was his start in microbe hunting!

3

He went back to London in 1894, plotting to throw up medicine and science. He was thirty-six. "Everything I had tried had failed," he wrote, but he consoled himself by imagining himself a sad defiant lone wolf: "But my failure did not depress me . . . it drove me aloft to peaks of solitude. . . . Such a spirit was a selfish spirit but nevertheless a high one. It desired nothing, it sought no praise . . . it had no friends, no fears, no loves, no hates."

But as you will see, Ronald Ross knew nothing of himself, for when he got going at his proper work, there was never a less calm and more desirous spirit than his. Nor a more enthusiastic one. And how he could hate!

When Ross returned to London he met Patrick Manson, an eminent and mildly famous English doctor. Manson had got himself medically notorious by discovering that mosquitoes can suck worms out of the blood of Chinamen (he had practiced in Shanghai); Manson had proved—this is remarkable!—that these worms can even develop in the stomachs of mosquitoes. Manson was obsessed by mosquitoes, he believed they were among the peculiar creatures of God, he was convinced they were important to the destinies of man, he was laughed at, and the medical wiseacres of Harley Street called him a "pathological Jules Verne." He was sneered at. And then he met Ronald Ross—whom the world had sneered at. What a pair of men these two were! Manson knew so little about mosquitoes that he believed they could only suck blood once in their lives, and Ross talked vaguely about mosquitoes and gnats not knowing that mosquitoes *were* gnats. And yet—

Manson took Ross to his office, and there he set Ross right about the malaria microbe of Laveran that Ross did not believe in. He showed Ronald Ross the pale malaria parasites, peppered with a blackish pigment. Together they watched these germs, fished out of the blood of sailors just back from the equator, turn into little squads of spheres inside the red blood cells, then burst out the blood cells. "That happens just when the man has his chill," explained Manson. Ross was amazed at the mysterious transformations and cavortings of the malaria germs in the blood. After those spheres had galloped out of the corpuscles, they turned suddenly into crescent shapes, then those crescents would shoot out two, three, four, sometimes six long whips, which lashed and curled about and made the beast look like a microscopic octopus.

"That, Ross, is the parasite of malaria—you never find it

in people without malaria—but the thing that bothers me is: How does it get from one man to another?"

Of course that didn't really bother Patrick Manson at all. Every cell in that man's brain had in it a picture of a mosquito or the memory of a mosquito or a speculation about a mosquito. He was a mild man, not a terrific worker himself, but intensely prejudiced on this subject of mosquitoes. And he appreciated Ronald Ross's energy of a dynamo, he knew Ronald Ross adored him, and he remembered Ross was presently returning to India. So one day, as they walked along Oxford Street, Patrick Manson took his jump: "Do you know, Ross," he said, "I have formed the theory that mosquitoes carry malaria . . . ?" Ronald Ross did not sneer or laugh.

Then the old doctor from Shanghai poured his fantastic theory over this young man whom he wanted to make his hands: "The mosquitoes suck the blood of people sick with malaria . . . the blood has those crescents in it . . . they get into the mosquito's stomach and shoot out those whips . . . the whips shake themselves free and get into the mosquito's carcass. . . . The whips turn into some tough form like the spore of an anthrax bacillus. . . . The mosquitoes die . . . they fall into water . . . people drink a soup of dead mosquitoes . . ."

This, mind you, was a story, a romance, a purely trumped-up guess on the part of Patrick Manson. But it was a passionate guess, and by this time you have learned, maybe, that one guess, guessed enthusiastically enough—one guess in a billion may lead to something in this strange game of microbe hunting. So this pair walked down Oxford Street. And Ross? Well, he talked about gnats and mosquitoes and did not know that mosquitoes were gnats. But Ross listened to Manson. . . . Mosquitoes carry malaria? That was an ancient superstition—but here was Doctor Manson, thinking about nothing else. Mosquitoes carry malaria? Well, Ross's books had not sold; his mathematics were ignored. . . . But here was a chance, a gamble! If Ronald Ross could prove mosquitoes were to blame for malaria! Why, a third of all the people in the hospitals in

India were in bed with malaria. More than a million a year died, directly or indirectly, because of malaria, in India alone! But if mosquitoes were really to blame—it would be easy!—malaria could be absolutely wiped out. . . . And if he, Ronald Ross, were the man to prove that!

"It is my duty to solve the problem," Ross said. Fictioneer that he was, he called it: "The Great Problem." And he threw himself at Manson's feet. "I am only your hands—it is your problem!" he assured the doctor from China.

"Before you go, you should find out something about mosquitoes," advised Manson, who himself didn't know whether there were ten different kinds of mosquitoes, or ten thousand, who thought mosquitoes could live only three days after they had bitten. So Ross (who didn't know mosquitoes were gnats) looked all over London for books about mosquitoes—and couldn't find any. Too little of a scholar, then, to think of looking in the library of the British Museum, Ross was sublimely ignorant, but maybe that was best, for he had nothing to unlearn. Never has such a green searcher started on such a complicated quest. . . .

He left his wife and children in England, and on the twenty-eighth of March, 1895, he set sail for India, with Patrick Manson's blessing, and full of his advice. Manson had outlined experiments—but how did one go about doing an experiment? But mosquitoes carry malaria! On with the mosquito hunt! On the ship Ross pestered the passengers, begging them to let him prick their fingers for a drop of blood. . . . He looked for mosquitoes, but they were not among the discomforts of the ship, so he dissected cockroaches—and he made an exciting discovery of a new kind of microbe in an unfortunate flying fish that had flopped on the deck. He was ordered to Secunderabad, a desolate military station that sat between hot little lakes in a huge plain dotted with horrid heaps of rocks, and here began to work with mosquitoes. He had to take care of patients too, he was only a doctor and the Indian Government—who can blame them?—would not for a moment recognize Ronald Ross as an official authentic microbe hunter or mosquito expert. He was alone. Everybody was against him from his colonel who thought him an insane upstart to the black-skinned boys who feared him for a dangerous nuisance (he was always wanting to prick their fingers!). The other doctors! They did not even believe in the malaria parasite. When they challenged him to show them the germs in the patient's blood, Ross went to the fray full of confidence, dragging after him a miserable Hindu whose blood was rotten with malaria microbes, but when the fatal test was made—curse it!—that wretched Hindu suddenly felt fit as a fiddle. His microbes had departed from him. The doctors roared with laughter. But Ronald Ross kept at it.

He started out to follow Manson's orders. He captured mosquitoes, any kind of mosquito, he couldn't for the life of him have told you what kind they were. He let the pests loose under nets over beds on which lay naked and foolishly superstitious dark-skinned people of a caste so low that they had no proper right to have emotions. The blood of these people was charmingly full of malaria microbes. The mosquitoes hummed under the nets—and wouldn't bite. Curse it! They could not be made to bite! "They are stubborn as mules," wrote Ross, in agony, to Patrick Manson. But he kept at it. He cajoled the mosquitoes. He pestered the patients. He put them in the hot sun "to bring their flavor out." The mosquitoes kept on humming and remained sniffish. But, eureka! At last he hit on the idea of pouring water over the nets, soaking the nets—also the patients, but that was no matter—and finally the mosquitoes got to work and sucked their fill of Hindu blood. Ronald Ross caught them then, put them gingerly in bottles, then day after day killed them and peeped into their stomachs to see if those malaria microbes they had sucked in with the blood might be growing. They didn't grow!

He bungled. He was like any tyro searcher—only his innate hastiness made him worse—and he was constantly making momentous discoveries that turned out not to be discoveries at all. But his bunglings had fire in them. To read his letters to Patrick Manson, you would think he had made himself miraculously small and crawled under the lens into that blood among the objects he was learning to spy upon. And what was best, everything was a story to him, no, more than a story, a melodrama. Manson had told him to watch those strange whips that grew out of the crescent malaria germs and made them look like octopuses. In vast excitement he wrote a long letter to Manson, telling of a strange fight between a whip that had shaken itself free, and a white blood cell—a phagocyte. He was a vivid man, was Ronald Ross. "He [Ross called that whip "he"] kept poking the phagocyte in the ribs (!) in different parts of his body, until the phagocyte finally turned and ran off howling . . . the fight between the whip and the phagocyte was wonderful. . . . I shall write a novel on it in the style of the 'Three Musketeers.' " That was the way he kept himself at it and got himself past the first ambushes and disappointments of his ignorance and inexperience. He collected malarious Hindus as a terrier collects rats. He loved them if they were shot full of malaria, he detested them when they got better. He gloried in the wretched Abdul Wahab, a dreadful case. He pounced on Abdul and dragged him from pillar to post. He put fleas on him. He tortured him with mosquitoes. He failed. He kept at it. He wrote to Manson: "Please send me advice. . . ." He missed important truths that lay right under his nose—that yelled to be discovered.

But he was beginning to know just exactly what a malaria parasite looked like—he could spot its weird black grains of pigment, and tell them apart from all of the unknown tiny blobs and bubbles and balloons that drifted before his eyes under his lens. And the insides of the stomachs of mosquitoes? They were becoming as familiar as the insides of this nasty hot quarters!

What an incredible pair of searchers they were! Away in London Patrick Manson kept answering Ross's tangled tortured letters, felt his way and gathered hope from his mixed-up accounts of unimportant experiments. "Let mosquitoes bite people sick with malaria," wrote Manson, "then put those mosquitoes in a bottle of water and let them lay eggs and hatch out grubs. Then give that mosquito-water to people to drink. . . ."

So Ross fed some of this malaria-mosquito soup to Lutchman, his servant, and almost danced with excitement as the man's temperature went up—but it was a false alarm, it wasn't malaria, worse luck. . . . So dragged the dreary days, the months, the years, feeding people mashed-up mosquitoes and writing to Manson: "I have a sort of feeling it will succeed— I feel a kind of religious excitement over it!" But it never succeeded. But he kept at it. He intrigued to get to places where he might find more malaria; he discovered strange new mosquitoes and from their bellies he dredged up unheard-of parasites—that had nothing to do with malaria. He tried everything. He was illogical. He was anti-scientific. He was like Edison combing the world to get proper stuff out of which to make phonograph needles. "There is only one method of solution," he wrote, "that is, by incessant trial and exclusion." He wrote that, while the simple method lay right under his hand, unfelt.

He wrote shrieking poems called "Wraths." He was ordered to Bangalore to try to stop the cholera epidemic, and didn't stop it. He became passionate about the Indian authorities. "I wish I might rub their noses in the filth and disease which they so impotently let fester in Hindustan," Ronald Ross cried. But who can blame him? It was hot there. "I was now forty years old," he wrote, "but, though I was well known in India, both for my sanitary work at Bangalore and for my researches on malaria I received no advancement at all for my pains."

4

So passed two years, until, in June of 1897 Ronald Ross came back to Secunderabad, to the steamy hospital of Begumpett. The monsoon bringing its cool rain should have already broken, but it had not. A hellish wind blew gritty clouds of dust into the laboratory of Ronald Ross. He wanted to throw his microscope out of the window. Its one remaining eyepiece was cracked, and its metal work was rusted with his sweat. There was the punka, the blessed punka, but he could not start the punka going because it blew his dead mosquitoes away, and in the evening when the choking wind had died, the dust still hid the sun in a dreadful haze. Ronald Ross wrote:

> What ails the solitude?
> Is this the judgment day?
> The sky is red as blood
> The very rocks decay.

And that relieved him and released him, just as another man might escape by whiskey or by playing bottle-pool, and on the sixteenth of August he decided to begin his work all over, to start, in short, where he had begun in 1895—"only much more thoroughly this time." So he stripped his malaria patient—it was the famous Husein Khan. Under the mosquito net went Husein, for Ronald Ross had found a new kind of mosquito with which to plague this Husein Khan, and in his unscientific classification Ross called this mosquito, simply, a brown mosquito. (For the purposes of historical accuracy, and to be fair to Battista Grassi, I must state that it is not clear where these brown mosquitoes came from. In the early part of his report Ronald Ross says he raised them from the grubs—but a moment later, speaking of a closely related mosquito, he says: "I have failed in finding their grubs also.")

It is no wonder—though lamentable for the purposes of history—that Ronald Ross was mixed up, considering his lone-wolf work and that hot wind and his perpetual failures! Anyway, he took those brown mosquitoes (which may have bitten other beasts, who knows) and loosed them out of their bottles under the net. They sucked the blood of Husein Khan, at a few cents per suck per mosquito, and then once more, one day after another, Ross peeped at the stomachs of those insects.

On the nineteenth of August he had only three of the brown beasts left. He cut one of them up. Hopelessly he began to look at the walls of its stomach, with its pretty, regular cells arranged like stones in a paved road. Mechanically he peered down the tube of his microscope, when suddenly something queer forced itself up into the front of his attention.

What was this? In the midst of the even pavement of the cells of the stomach wall lay a funny circular thing, about a twenty-five-hundredth of an inch its diameter was—here was another! But, curse it! It was hot—he stopped looking. . . .

The next day it was the same. Here, in the wall of the stomach of the next to the last mosquito, four days after it had sucked the blood of the unhappy malarious Husein Khan, here were those same circular outlines—clear—much more distinct than the outlines of the cells of the stomach, and in each one of these circles was "a cluster of small granules, black as jet!" Here was another of those fantastic things, and another—he counted twelve in all. He yawned. It was hot. That black pigment looked a lot like the black pigment inside of malaria microbes in the blood of human bodies—but it was hot. Ross yawned, and went home for a nap.

And as he awoke—so he says in his memoirs—a thought struck him: "Those circles in the wall of the stomach of the mosquito—those circles with their dots of black pigment, they can't be anything else than the malaria parasite, growing there. . . . That black pigment is just like the specks of black pigment in the microbes in the blood of Husein Khan. . . . The longer I wait to kill my mosquitoes after they have sucked his blood, the bigger those circles should grow . . . if they are alive, they *must* grow!"

Ross fidgeted about—and how he could fidget!—waiting for the next day, that would be the fifth day after his little flock of mosquitoes had fed on Husein under the net. That was the day for the cutting up of the last mosquito of the flock. Came the twenty-first of August. "I killed my last mosquito," Ronald Ross wrote to Manson, "and rushed at his stomach!"

Yes! Here they were again, those circle cells, one . . . two . . . six . . . twenty of them. . . . They were full of the same jet-black dots. . . . Sure enough! They were bigger than the circles in the mosquito of the day before. . . . They were really growing! They *must* be the malaria parasites growing! (Though there was no absolutely necessary reason they must be.) But they must be! Those circles with their black dots in the bellies of three measly mosquitoes now kicked Ronald Ross up to heights of exultation. He must write verses!

> I have found thy secret deeds
> Oh, million-murdering death.

> I know that this little thing
> A million men will save—
> Oh, death, where is thy sting?
> Thy victory, oh, grave?

At least that is what Ronald Ross, in those memoirs of his, says he wrote on the night of the day of his first little success. But to Manson, telling the finest details about the circles with their jet-black dots, he only said:

"The hunt is up again. It may be a false scent, but it smells promising."

And in a scientific paper, sent off to England to the *British Medical Journal*, Ronald Ross wrote gravely like any cool searcher. He wrote admitting he had not taken pains to study his brown mosquitoes carefully. He admitted the jet-black dots might not be malaria parasites at all, but only pigment coming from the blood in the mosquito's gullet. There certainly was need for this caution, for he was not sure where his brown mosquitoes came from: some of them might have sneaked in through a hole in the net—and those intruders *might* have bitten a bird or beast before they fed on his Hindu patient. It was a most mixed-up business. But he could write poems about saving the lives of a million men!

Such a man was Ronald Ross, mad poet shaking his fist in the face of the malignant Indian sun, celebrating uncertain discoveries with triumphant verses, spreading nets with maybe no holes in them. . . . But you must give him this: he had been lifted up. And, as you will see, it was to the everlasting honor of Ronald Ross that he was exalted by this seemingly so piffling experiment. He clawed his way—and this is one of the major humors of human life!—with unskilled but enthusiastic fingers toward the uncovering of a murderous fact and a complicated fact. A fact you would swear it would take the sure intelligence of some god to uncover.

Then came one of those deplorable interludes. The High Authorities of the Indian Medical Service failed to appreciate him. They sent him off to active duty at doctoring, mere doctoring. Ronald Ross rained telegrams on his Principal Medical Officer. He implored Manson way off there in England. In vain. They packed him off up north, where there were few mosquitoes, where the few he did catch would not bite—it was so cold, where the natives (they were Bhils) were so superstitious and savage they would not let him prick their fingers. All he could do was fish trout and treat cases of itch. How he raved!

5

But Patrick Manson did not fail him, and presently Ross came down from the north, to Calcutta, to a good laboratory, to assistants, to mosquitoes, to as many—for that city was a fine malaria pest-hole!—Hindus with malaria crescents in their blood as any searcher could possibly want. He advertised for helpers. An assorted lot of dark-skinned men came, and of these he chose two. The first, Mahomed Bux, Ronald Ross hired because he had the appearance of a scoundrel, and (said Ross) scoundrels are much more likely to be intelligent. The second assistant Ross chose was Purboona. All we know of that man is that he had the booming name of Purboona, and Purboona lost his chance to become immortal because he vamoosed after his first pay day.

So Ross and Mr. Mahomed Bux set to work to try to find once more the black-dotted circles in the stomachs of mosquitoes. Mr. Mahomed Bux sleuth-footed it about, among the sewers, the drains, the stinking tanks of Calcutta, catching gray mosquitoes and brindled mosquitoes and brown and green dappled-winged ones. They tried all kinds of mosquitoes (within the limits of Ronald Ross's feeble knowledge of the existing kinds). And Mr. Mahomed Bux? He was a howling success. The mosquitoes seemed to like him, they would bite Hindus for this wizard of a Mahomed when Ross could not make them bite at all—Mahomed whispered things to his mosquitoes. . . . And a rascal? No. Mr. Mahomed Bux had just one little weakness—he faithfully got thoroughly drunk once a week on *Ganja*. But the experiments? They turned out as miserably as Mahomed turned out beautifully, and it was easy for Ross to wonder whether the heat was causing him to see things last year at Begumpett.

Then the God of Gropers came to help Ronald Ross. Birds have malaria. The malaria microbe of birds looks very like the malaria microbe of men. Why not try birds?

So Mr. Mahomed Bux went forth once more and cunningly snared live sparrows and larks and crows. They put them in cages, on beds, with mosquito bar over the cages, and Mahomed slept, with one eye open, on the floor between the beds to keep away the cats.

On St. Patrick's day of the year 1898, Ronald Ross let loose ten gray mosquitoes into a cage containing three larks, and the blood of those larks teemed with the germs of malaria. The ten mosquitoes bit those larks, and filled themselves with lark's blood.

Three days later Ronald Ross could shout: "The microbe of the malaria of birds grows in the wall of the stomach of the gray mosquito—just as the human microbe grew in the wall of the stomach of the brown spot-winged mosquito."

Then he wrote to Patrick Manson. This lunatic Ross became for a moment himself a malaria microbe! That night he wrote these strange words to Patrick Manson:

"I find that I exist constantly in three out of four mosquitoes fed on bird-malaria parasites, and that I increase regularly in size from about a seven-thousandth of an inch after about thirty hours to about one seven-hundredth of an inch after about eighty-five hours. . . . I find myself in large numbers in about one out of two mosquitoes fed on two crows with blood parasites. . . ."

He thought he was himself a circle with those jet-black dots. . . .

"What an ass I have been not to follow your advice before and work with birds!" Ross wrote to Manson. Heaven knows what Ronald Ross would have discovered without that persistent Patrick Manson.

You would think that such a man as Ross, wild as the maddest of hatters, topsy-turvy as the dream of a hasheesh-eater, you would swear, I say, that he could do no accurate experiments. Wrong! For presently he was up to his ears in an experiment Pasteur would have been proud to do.

Mr. Mahomed Bux brought in three sparrows, and one of these sparrows was perfectly healthy, with no malaria microbes in its blood; the second had a few; but the third sparrow was very sick—his blood swarmed with the black-dotted germs. Ross took these three birds and put each one in a separate cage, mosquito-proof. Then the artful Mahomed took a brood of she-mosquitoes, clean, raised from the grubs, free of all suspicion of malaria. He divided this flock up into three little flocks, he whispered Hindustani words of encouragement to them. Into each cage, with its sparrow, he let loose a flock of these mosquitoes.

Marvelous! Not a mosquito who sucked the blood of the healthy sparrow showed those dotted circles in her stomach. The insects who had bitten the mildly sick bird had a few. And Ronald Ross, peeping through his lens at the stomachs of the mosquitoes who had bitten the very sick sparrow—

found their gullets fairly polka-dotted with the jet-black pigmented circles!

Day after day Ross killed and cut up one after another of the last set of mosquitoes. Day after day, he watched those circles swelling, growing—there was no doubt about it now; they began to look like warts sticking out of the wall of the stomach. And he watched weird things happening in those warts. Little bright colored grains multiplied in them, "like bullets in a bag." Were these young malaria microbes? Then where did they go from here? How did they get into new healthy birds? Did they, indeed, get from mosquitoes into other birds?

Excitedly Ronald Ross wrote to Patrick Manson: "Well, the theory is proved, the mosquito theory is a fact." Which of course it wasn't, but that was the way Ronald Ross encouraged himself. There was another regrettable interlude, in which the unseen hand of his incurable restless dissatisfaction took him by the throat, and dragged him away up north to Darjeeling, to the hills that make giant's steps up to the white Himalayas, but of this interlude we shall not speak, for it was lamentable, this restlessness of Ronald Ross, with the final simple experiment fairly yelling to be done. . . .

But by the beginning of June he was back at his birds in Calcutta—it was more than 100 degrees in his laboratory— and he was asking: "Where do the malaria microbes go from the circles that grow into those big warts in the stomach wall of the mosquito?"

They went, those microbes, to the spit-gland of those mosquitoes!

Squinting through his lens at a wart on the wall of the stomach of a she-mosquito, seven days after she had made a meal from the blood of a malarious bird, Ronald Ross saw that wart burst open! He saw a great regiment of weird spindle-shaped threads march out of that wart. He watched them swarm through the whole body of that she-mosquito. He pawed around in countless she-mosquitoes who had fed on malarious birds. He watched other circles grow into warts, get ripe, burst, shoot out those spindles. He pried through his lens at the "million things that go to make up a mosquito"—he hadn't the faintest notion what to call most of them—until one day, strangest of acts of malignant nature, he saw those regiments of spindle-threads, which had teemed in the body of the mosquito, march to her spit-gland.

In that spit-gland, feebly, lazily moving in it, but swarming in such myriads that they made it quiver, almost, under his lens, were those regiments and armies of spindle-shaped threads, hopeful valiant young microbes of malaria, ready to march up the tube to the mosquito's stinger. . . .

"It's by the bite mosquitoes carry malaria then," Ross whispered—he whispered it because that was contrary to the theory of his scientific father, Patrick Manson. "It is all nonsense that birds—or people either—get malaria by drinking dead mosquitoes, or by inhaling the dust of mosquitoes. . . ." Ronald Ross had always been loyal to Patrick Manson. But now! Never has there been a finer instance of wrong theories leading a microbe hunter to unsuspected facts. But now! Ronald Ross needed no help. He was a searcher.

"It's by the bite!" shouted Ronald Ross, so, on the twenty-fifth day of June in 1898, Mr. Mahomed Bux brought in three perfectly healthy sparrows—fine sparrows with not a single microbe of malaria in their blood. That night, and night after night after that night, with Ronald Ross watching, Mr. Mahomed Bux let into the cage with those healthy sparrows a flock of poisonous she-mosquitoes who had fed on sick birds. . . . And Ronald Ross, fidgety as a father waiting news of his first-born child, biting his mustache, sweating, and sweating more yet because he used up so much of himself cursing at his sweat—Ross watched those messengers of death bite the healthy sparrows. . . .

On the ninth of July Ross wrote to Patrick Manson: "All three birds, perfectly healthy before, are now simply swarming with proteosoma." (Proteosoma are the malarial parasites of birds.)

Now Ronald Ross did anything but live remotely on his mountain top. He wrote this to Manson, he wired it to Manson, he wrote it to Paris to old Alphonse Laveran, the discoverer of the malaria microbe; he sent papers to one scientific journal and two medical journals about it; he told everybody in Calcutta about it; he bragged about it—in short, this Ronald Ross was like a boy who had just made his first kite finding that the kite could really fly. He went wild—and then (it is too bad!) he collapsed. Patrick Manson went to Edinburgh and told the doctors of the great medical congress about the miracle of the sojourn and the growing and the meanderings of the malaria microbes in the bodies of gray she-mosquitoes: he described how his protégé Ronald Ross, alone, obscure, laughed at, but tenacious, had tracked the germ of malaria from the blood of a bird through the belly and body of she-mosquitoes to their dangerous position in her stinger, ready to be shot into the next bird she bit.

The learned doctors gaped. Then Patrick Manson read out a telegram from Ronald Ross. It was the final proof: the bite of a malarial mosquito had given a healthy bird malaria! The congress—this is the custom of congresses—permitted itself a dignified furore, and passed a resolution congratulating this unknown Major Ronald Ross on his "Great and Epoch-Making Discovery." The congress—it is the habit of congresses—believed that what is true for birds goes for men too. The congress—men in the mass are ever uncritical—thought that this meant malaria would be wiped out from to-morrow on and forever—for what is simpler than to kill mosquitoes? So that congress permitted itself a furore.

But Patrick Manson was not so sure: "One can object that the facts determined for birds do not hold, necessarily, for men." He was right. There was the rub. This was what Ronald Ross seemed to forget: that nature is everlastingly full of surprises and annoying exceptions, and if there are laws and rules for the movements of the planets, there may be absolutely no apparent rime and less reason for the meanderings of the microbes of malaria. . . . Searchers, the best of them, still do no more than scratch the surface of the most amazing mysteries, all they can do (yet!) to find truth about microbes is to hunt, hunt endlessly. . . . There are no laws!

So Patrick Manson was stern with Ronald Ross. This nervous man, feeling he could stand this cursed India not one moment longer, must stand it months longer, years longer! He had made a brilliant beginning, but only a beginning. He must keep on, if not for science, or for himself, then for England! For England! And in October Manson wrote him: "I hear Koch has failed with the mosquito in Italy, so you have time to grab the discovery for England."

But Ronald Ross—alas—could not grab that discovery of *human* malaria, not for science, nor humanity, nor for England—nor (what was worst) for himself. He had come to the end of his rope. And among all microbe hunters, there is for me no more tortured man than this same Ronald Ross. There have been searchers who have failed—they have kept on hunting with the naturalness of ducks swimming; there have been searchers who have succeeded gloriously—but they were hunters born, and they kept on hunting in spite of the seductions of glory. But Ross! Here was a man who could only do patient experiments—with a tragic impatience, in agony, against the clamoring of his instincts that yelled against the priceless loneliness that is the one condition for all true searching. He had visions of himself at the head of important committees, and you can *feel* his dreams of medals and banquets and the hosannahs of multitudes. . . .

He must grab the discovery for England. He tried gray mosquitoes and green and brown and dappled-winged mosquitoes on Hindus rotten with malaria—but it was no go! He became sleepless and lost eleven pounds. He forgot things. He could not repeat even those first crude experiments at Secunderabad.

And yet—all honor to Ronald Ross. He did marvelous things in spite of himself. It was his travail that helped the learned, the expert, the indignant Battista Grassi to do those clean superb experiments that must end in wiping malaria from the earth.

6

You might know Giovanni Battista Grassi would be the man to do what Ronald Ross had not quite succeeded in bringing off. He had been educated for a doctor, at Pavia where that glittering Spallanzani had held forth amid applause a hundred years before. Grassi had been educated for a doctor (Heaven knows why) because he had no sooner got his license than he set himself up in business as a searcher in zoölogy. With a certain amount of sniffishness he always insisted: "I am a zoologo—not a medico!" Deliberate as a glacier, precise as a ship's chronometer, he started finding answers to the puzzles of nature. Correct answers! His works were pronounced classics right after he published them—but it was his habit not to publish them for years after he started to do them. He made known the secret comings and goings of the Society of the White Ants—not only this, but he discovered microbes that plagued and preyed upon these white ants. He knew more than any man in the world about eels—and you may believe it took a searcher with the insight of a Spallanzani to trace out the weird and romantic changes that eels undergo to fulfill their destiny as eels. Grassi was not strong. He had abominable eyesight. He was full of an argumentative petulance. He was a contradictory combination of a man too modest to want his picture in the papers but bawling at the same time for the last jot and tittle of credit for everything that he did. And he did everything. Already, when he was only twenty-nine, before Ross had dreamt of becoming a searcher, Battista Grassi was a professor, and had published his famous monograph upon the Chaetognatha (I do not know what they are!).

Before Ronald Ross knew that anybody had ever thought of mosquitoes carrying malaria, Grassi had had the idea, had taken a whirl at experiments on it, but had used the wrong mosquito, and failed. But that failure started ideas stewing in his head while he worked at other things—and how he worked! Grassi detested people who didn't work. "Mankind," he said, "is composed of those who work, those who pretend to work, and those who do neither." He was ready to admit that he belonged in the first class, and it is entirely certain that he did belong there.

In 1898, the year of the triumph of Ronald Ross, Grassi, knowing nothing of Ross, never having heard of Ross, went back at malaria again. "Malaria is the worst problem Italy has to face! It desolates our richest farms! It attacks millions in our lush lowlands! Why don't you solve that problem?" So the politicians, to Battista Grassi. Then too, the air was full of whispers of the possibility that I don't know how many different diseases might be carried from man to man by insects. There was that famous work of Theobald Smith, and Grassi had an immense respect for Theobald Smith. But what probably finally set Grassi working at malaria—you must remember he was a very patriotic and jealous man—was the arrival of Robert Koch. Dean of the microbe hunters of the world, Tsar of Science (his crown was only a little battered) Koch had come to Italy to prove that mosquitoes carry malaria from man to man.

Koch was an extremely grumpy, quiet, and restless man now; sad because of the affair of his consumption cure (which had killed a considerable number of people); restless after the scandal of his divorce from Emmy Fraatz. So Koch went from one end of the world to the other, offering to conquer plagues but not quite succeeding, trying to find happiness and not quite reaching it. His touch faltered a little. . . . And now Koch met Battista Grassi, and Grassi said to Robert Koch:

"There are places in Italy where mosquitoes are absolutely pestiferous—but there is no malaria at all in those places!"

"Well—what of it?"

"Right off, that would make you think mosquitoes had nothing to do with malaria," said Battista Grassi.

"So?" ... Koch was enough to throw cold water on any logic!

"Yes—but here is the point," persisted Grassi, "I have not found a single place where there is malaria—where there aren't mosquitoes too!"

"What of that?"

"This of that!" shouted Battista Grassi. "Either malaria is carried by one special particular blood-sucking mosquito, out of the twenty or forty kinds of mosquitoes in Italy—or it isn't carried by mosquitoes at all!"

"Hrrrm-p," said Koch.

So Grassi made no hit with Robert Koch, and so Koch and Grassi went their two ways, Grassi muttering to himself: "Mosquitoes—without malaria ... but never malaria—without mosquitoes! That means one special kind of mosquito! I must discover the suspect. . . ."

That was the homely reasoning of Battista Grassi. He compared himself to a village policeman trying to discover the criminal in a village murder. "You wouldn't examine the whole population of a thousand people one by one!" muttered Grassi. "You would try to locate the suspicious rogues first. . . ."

His lectures for the year 1898 at the University of Rome over, he was a conscientious man who always gave more lectures than the law demanded, he needed a rest, and on the 15th of July he took it. Armed with sundry fat test-tubes and a notebook, he sallied out from Rome to those low hot places and marshy desolations where no man but an idiot would go for a vacation. Unlike Ross, this Grassi was a mosquito expert besides everything else that he was. His eyes—so red-rimmed and weak—were exceedingly sharp at spotting every difference between the thirty-odd different kinds of mosquitoes that he met. He went around with the fat test-tube in his hand, his ear cocked for buzzes. The buzz dies away as the mosquito lights. She has lit in an impossible place. Or she has lit in a disgusting place. No matter, Battista Grassi is up behind her, pounces on her, claps his fat test-tube over her, puts a grubby thumb over the mouth of the test-tube, paws over his prize and pulls her apart, scrawls little cramped pothooks in his notebook. That was Battista Grassi, up and down and around the nastiest places in Italy all that summer.

So it was he cleared a dozen or twenty different mosquitoes of the suspicion of the crime of malaria—he was always finding these beasts in places where there was no malaria. He ruled out two dozen different kinds of gray mosquitoes and brindled mosquitoes, that he found anywhere—in saloons and bedrooms and the sacristies of cathedrals, biting babies and nuns and drunkards. "You are innocent!" shouted Battista Grassi at these mosquitoes. "For where you are none of these nuns or babies or drunkards suffers from malaria!"

You will grant this was a most outlandish microbe hunting of Grassi's. He went around making a nuisance of himself. He insinuated himself into the already sufficiently annoyed families of those hot malarious towns. He snooped annoyingly into the affairs of these annoyed families: "Is there malaria in your house? . . . Has there ever been malaria in your house? . . . How many have never had malaria in your house . . . how many mosquito bites did your sick baby have last week? . . . What kind of mosquitoes bit him?" He was utterly without a sense of humor. And he was annoying.

"No," the indignant head of the house might tell him, "we suffer from malaria—but we are not bothered by mosquitoes!" Battista Grassi would never take his word for that. He snouted into pails and old crocks in the back yards. He peered beneath tables and behind sacred images and under beds. He even discovered mosquitoes hiding in shoes under those beds. . . .

So it was—it is most fantastical—that Battista Grassi went more than two-thirds of the way to solving this puzzle of how malaria gets from sick men to healthy ones before he had ever made a single experiment in his laboratory! For, everywhere where there was malaria, there *were* mosquitoes. And *such* mosquitoes! They were certainly a very special definite sort of blood-sucking mosquito Grassi found.

"Zan-za-ro-ne, we call that kind of mosquito," the house-holders told him.

Always, where the "zan-za-ro-ne" buzzed, there Grassi found deep flushed faces on rumpled beds, or faces with chattering teeth going towards those beds. Always where that special and definite mosquito sang at twilight, Grassi found fields waiting for some one to till them, and from the houses of the little villages that sat in these fields, he saw processions emerging, and long black boxes. . . .

There was no mistaking this mosquito, zanzarone, once you had spotted her; she was a frivolous gnat that flew up from the marshes towards the lights of the towns; she was an elegant mosquito proud of four dark spots on her light brown wings; she was not a too dignified insect who sat in an odd way with the tail-end of her body sticking up in the air [that was one way he could spot her, for the Culex mosquitoes drooped their tails]; she was a brave blood-sucker who thought: "The bigger they are the more blood I get out of them!" So zanzarone preferred horses to men and men to rabbits. That was zanzarone, and the naturalists had given her the name *Anopheles claviger* many years before. *Anopheles claviger!* This became the slogan of Battista Grassi. You can see him, shuffling along behind lovers in the dusk, making fists of his fingers to keep himself from pouncing on the zanzarone who made meals off their regardless necks. . . . You can see this Grassi, sitting in a stagecoach with no springs, oblivious to bumps, deaf to the chatter of his fellow-passengers, with absent eyes counting the *Anopheles claviger* he had discovered—with delight—riding on the ceiling of the wagon in which he journeyed from one utterly terrible little malarious village to another still more cursed.

"I'll try them on myself!" Grassi cried. He went up north to his home in Rovellasca. He taught boys how to spot the anopheles mosquito. The boys brought boxes full of these she-zanzarone from towns where malaria raged. Grassi took these boxes to his bedroom, put on his night shirt, opened the boxes, crawled into bed—but curse it! not one of the zanzarone bit

him. Instead they flew out of his room and bit Grassi's mother, "fortunately without ill effect!"

Then Grassi went back to Rome to his lectures, and on September 28th of 1898, before ever he had done a single serious experiment, he read his paper before the famous and ancient Academy of the Lincei: "It is the anopheles mosquito that carries malaria if any mosquito carries malaria . . ." And he told them he was suspicious of two other brands of mosquitoes—but that was absolutely all, out of the thirty or forty different tribes that infected the low places of Italy.

Then came an exciting autumn for Battista Grassi and an entertaining autumn for the wits of Rome, and a most important autumn for mankind. Besides all that it was a most itchy autumn for Mr. Sola, who for six years had been a patient of Dr. Bastianelli in the Hospital of the Holy Spirit, high up on the top floor of this hospital that sat on a high hill of Rome. Here zanzarone never came. Here nobody ever got malaria. Here was the place for experiments. And here was Mr. Sola, who had never had malaria, every twist and turn of whose health Dr. Bastianelli knew, who told Battista Grassi that he would not mind being shut up with three different brands of hungry she-mosquitoes every night for a month.

Grassi and Bignami and Bastianelli started off, strangely enough, with those two minor mosquito suspects—those two culexes that Grassi had discovered always hanging around malarious places along with the zanzarone. . . . They tortured Mr. Sola each night with hundreds of these mosquitoes. They shut poor Mr. Sola up in that room with those devils and turned off the light. . . .

Nothing happened. Sola was a tough man. Sola showed not a sign of malaria.

(It is not clear why Grassi did not start off by loosing his zanzarone at this Mr. Sola.)

Maybe it was because Robert Koch had laughed publicly at this idea of the zanzarone—Grassi does admit that discouraged him.

But, one fine morning, Grassi hurried out of Rome to Moletta and came back with a couple of little bottles in which buzzed ten fine female anopheles mosquitoes. That night Mr. Sola had a particularly itchy time of it. Ten days later this stoical old gentleman shook horribly with a chill, his body temperature shot up into a high fever—and his blood swarmed with the microbes of malaria.

"The rest of the history of Sola's case has no interest for us," wrote Grassi, "but it is now certain that mosquitoes can carry malaria, to a place where there are no mosquitoes in nature, to a place where no case of malaria has ever occurred, to a man who has never had malaria—Mr. Sola!"

Over the country went Grassi once more, chasing zanzarone, hoarding zanzarone: in his laboratory he tenderly raised zanzarone on winter-melons and sugar-water; and in the top of the hospital of the Holy Spirit, in those high mosquito-proof rooms, Grassi and Bastianelli (to say nothing of another assistant, Bignami) loosed zanzarone into the bedrooms of people who had never had malaria—and so gave them malaria.

It was an itchy autumn and an exciting one. The newspapers became sarcastic and hinted that the blood of these poor human experimental animals would be on the heads of these three conspirators. But Grassi said: to the devil with the newspapers, he cheered when his human animals got sick, he gave them doses of quinine as soon as he was sure his zanzarone had given them malaria, and then "their histories had no further interest for him."

By now Grassi had read of those experiments of Ronald Ross with birds. "Pretty crude stuff!" thought this expert Grassi, but when he came to look for those strange doings of the circles and warts and spindle-shaped threads in the stomachs and saliva-glands of his she-anopheles, he found that Ronald Ross was exactly right! The microbe of human malaria in the body of his zanzarone did exactly the same things the microbe of bird malaria had done in the bodies of those mosquitoes Ronald Ross hadn't known the names of. Grassi didn't waste too much time praising Ronald Ross, who, Heaven knows, deserved praise, needed praise, and above all *wanted* praise. Not Grassi!

"By following my own way I have discovered that a special mosquito carried human malaria!" he cried, and then he set out—"It is with great regret I do this," he explained—to demolish Robert Koch. Koch had been fumbling and muddling. Koch thought malaria went from man to man just as Texas fever traveled from cow to cow. Koch believed baby mosquitoes inherited malaria from their mothers, bit people, and so infected them. And Koch had sniffed at the zanzarone.

So Grassi raised baby zanzarone. He let them hatch out in a room, and every evening in this room, for four months, sat this Battista Grassi with six or seven of his friends. What friends he must have had! For every evening they sat there in the dusk, barelegged with their trousers rolled up to their knees, bare-armed with their shirt sleeves rolled up to their elbows. Some of these friends, whom the anopheles relished particularly, were stabbed every night fifty or sixty times! So Grassi demolished Robert Koch, and so he proved his point, because, though the baby anopheles were children of mother mosquitoes who came from the most pestiferous malaria holes in Italy, not one of Grassi's friends had a sign of malaria!

"It is not the mosquito's children, but only the mosquito who herself bites a malaria sufferer—it is only that mosquito who can give malaria to healthy people!" cried Grassi.

Grassi was as persistent as Ronald Ross had been erratic. He plugged up every little hole in his theory that anopheles is the one special and particular mosquito to bring malaria to men. By a hundred air-tight experiments he proved the malaria of birds could not be carried by the mosquitoes who brought it to men and that the malaria of men could never be strewn abroad by the mosquitoes who brought it to birds. Nothing was too much trouble for this Battista Grassi! He knew as much about the habits and customs and traditions of those zanzarone as if he himself were a mosquito and the king and ruler of mosquitoes. . . .

7

What is more, Battista Grassi was a practical man, and as I have said, an excessively patriotic man. He wanted to see his discovery do well by Italy, for he loved his Italy faithfully and violently. His experiments were no sooner finished, the last good strong nail was no sooner driven into the house of his case against the anopheles, than he began telling people, and writing in newspapers, and preaching—you might almost say he went about, bellowing till he bored everybody:

"Keep away the zanzarone and in a few years Italy will be free from malaria!"

He became a fanatic on the best ways to kill anopheles: he was indignant (that man had no sense of humor!) because townspeople insisted on strolling through their streets in the dusk. "How can you be so foolish as to walk in the twilight?" Grassi asked them. "That is the very time when the malaria mosquito is waiting for you."

He was the very type of the silly sanitarian. "Don't go out in the warm evenings," he told every one, "unless you wear heavy cotton gloves and veils!" (Imagine young Italians making love in heavy cotton gloves and veils.) So there was a good deal of sniggering at this professor who had become a violent missionary against the zanzarone.

But Battista Grassi was a practical man! "One family, staying free from the tortures of malaria—that would be worth ten years of preaching—I'll have to *show* them!" he muttered. So, in 1900, after his grinding experiments of 1898 and '99, this tough man set out to "show them." He went down into the worst malaria region of Italy, along the railroad line that ran through the plain of Capaccio. It was high summer. It was deadly summer there, and every summer the poor wretches of railroad workers, miserable farmers whose blood was gutted by the malaria poison, would leave that plain, at the cost of their jobs, at the cost of food, at the risk of starvation—to the hills to flee the malaria. And every summer from the swamps at twilight swarmed the malignant hosts of female zanzarone; at each hot dusk they made their meals and did their murders, and in the night, bellies full of blood, they sang back to their marshes, to marry and lay eggs and hatch out thousands more of their kind.

In the summer of 1900 Battista Grassi went to the plain of Capaccio. The hot days were just beginning, the anopheles were on the march. In the windows and on the doors of ten little houses of station-masters and employees of the railroad Grassi put up wire screens, so fine-meshed and so perfect that the slickest and the slightest of the zanzarone could not slip through them. Then Grassi, armed with authority from the officials of the railroad, supplied with money by the Queen of Italy, became a task-master, a Pharaoh with lashes. One hundred and twelve souls—railroad men and their families—became the experimental animals of Battista Grassi and had to be careful to do as he told them. They had to stay indoors in the beautiful but dangerous twilight. Careless of death—especially unseen death—as all healthy human beings are careless, these one hundred and twelve Italians had to take precautions, to avoid the stabs of mosquitoes. Grassi had the devil of a time with them. Grassi scolded them. Grassi kept them inside those screens by giving them prizes of money. Grassi set them an indignant example by coming down to Albanella, most deadly place of all, and sleeping two nights a week behind those screens.

All around those screen-protected station houses the zanzarone swarmed in humming thousands—it was a frightful year for mosquitoes. Into the *un-screened* neighboring station houses (there were four hundred and fifteen wretches living in those houses), the zanzarone swooped and sought their prey. Almost to a man, woman, and child, those four hundred and fifteen men, women and children fell sick with the malaria.

And of those one hundred and twelve prisoners behind the screens at night? They were rained on during the day, they breathed that air that for a thousand years the wisest men were sure was the cause of malaria, they fell asleep at twilight, they did all of the things the most eminent physicians had always said it was dangerous to do, but in the dangerous evenings they stayed behind screens—and only five of them got the malaria during all that summer. Mild cases these were, too, maybe only relapses from the year before, said Grassi.

"In the so-much-feared station of Albanella, from which for years so many coffins had been carried, one could live as healthily as in the healthiest spot in Italy!" cried Grassi.

8

Such was the fight of Ronald Ross and Battista Grassi against the assassins of the red blood corpuscles, the sappers of vigorous life, the destroyer of men, the chief scourge of the lands of the South—the microbe of malaria. There were aftermaths of this fight, some of them too long to tell, and some too painful. There were good aftermaths and bad ones. There are fertile fields now, and healthy babies, in Italy and Africa and India and America, where once the hum of the anopheles brought thin blood and chattering teeth, brought desolate land and death.

There is the Panama Canal. . . .

Then there is Sir Ronald Ross, who was—as once he hoped and dreamed—given enthusiastic banquets.

There is Ronald Ross who got the Nobel Prize of seven thousand eight hundred and eighty pounds sterling for his discovery of how the gray mosquito carries malaria to birds. . . .

There is Battista Grassi who didn't get the Nobel Prize, and is now unknown, except in Italy, where they huzzahed for him and made him a Senator (he never missed a meeting of that Senate to within a year of his death).

All these are, for the most part, good, even if some of them are slightly ironical aftermaths.

Then there is Ronald Ross, who had learned the hard game of searching while he made his discovery about the gray mosquito—you would say his best years of work were just beginning—there is Ronald Ross, insinuating Grassi was a thief, hinting that Grassi was a charlatan, saying Grassi had added almost nothing to the proof that mosquitoes carry malaria to men!

There was Grassi—justifiably purple with indignation, writing violent papers in reply. . . . You cannot blame him! But why will such searchers scuffle, when there are so many things left to find? You would think—of course it would be so in a novel—that they could have ignored each other, or could have said: "The facts of science are greater than the little men who find those facts!"—and then have gone on searching, and saving.

For the fight has only just begun. The day I finish this tale, it is twenty-five years after the perfect experiment of Grassi, comes this news item from Tokio—it is stuck away down in a corner of an inside page of a newspaper:

"The population of the Ryukyu Islands, which lie between Japan and Formosa, is rapidly dying off. . . . Malaria is blamed principally. In eight villages of the Yaeyama group . . . not a single baby has been born for the last thirty years. In Nozoko village . . . one sick old woman was the only inhabitant. . . ."

11

Walter Reed

In the Interest of Science—and for Humanity!

1

With yellow fever it was different—there were no brawls about it.

Everybody is agreed that Walter Reed—head of the Yellow Fever Commission—was a courteous man and a blameless one, that he was a mild man and a logical: there is not one particle of doubt he had to risk human lives; animals simply will not catch yellow fever!

Then it is certain that the ex-lumberjack, James Carroll, was perfectly ready to let go his own life to prove Reed's point, and he was not too sentimental about the lives of others when *he* needed to prove a point—which might and might not be what you would call a major point.

All Cubans (who were on the spot and ought to know) are agreed that those American soldiers who volunteered for the fate of guinea-pigs were brave beyond imagining. All Americans who were then in Cuba are sure that those Spanish immigrants who volunteered for the fate of guinea-pigs were not brave, but money-loving—for didn't each one of them get two hundred dollars?

Of course you might protest that fate hit Jesse Lazear a hard knock—but it was his own fault: why didn't he brush that mosquito off the back of his hand instead of letting her drink her fill? Then, too, fate has been kind to his memory; the United States Government named a Battery in Baltimore Harbor in his honor! And that same government has been more than kind to his wife: the widow Lazear gets a pension of fifteen hundred dollars a year! You see, there are no arguments—and that makes it fun to tell this story of yellow fever. And aside from the pleasure, it has to be told: this history is absolutely necessary to the book of Microbe Hunters. It vindicates Pasteur! At last Pasteur, from his handsome tomb in that basement in Paris, can tell the world: "I told you so!" Because, in 1926, there is hardly enough of the poison of yellow fever left in the world to put on the points of six pins; in a few years there may not be a single speck of that virus left on earth—it will be as completely extinct as the dinosaurs— unless there is a catch in the fine gruesome experiments of Reed and his Spanish immigrants and American soldiers. . . .

It was a grand coöperative fight, that scotching of the yellow jack. It was fought by a strange crew, and the fight was begun by a curious old man, with enviable mutton chop whiskers—his name was Doctor Carlos Finlay—who made an amazingly right guess, who was a terrible muddler at experiments, who was considered by all good Cubans and wise doctors to be a Theorizing Old Fool. What a crazy crank is Finlay, said everybody.

For everybody knew just how to fight that most panic-striking plague, yellow fever; everybody had a different idea of just how to combat it. You should fumigate silks and satins and possessions of folks before they *left* yellow fever towns— no! that is not enough: you should burn them. You should bury, burn, and utterly destroy these silks and satins and possessions before they *come into* yellow fever towns. It was wise not to shake hands with friends whose families were dying of yellow fever; it was perfectly safe to shake hands with them. It was best to burn down houses where yellow fever had lurked—no! it was enough to smoke them out with sulphur. But there was one thing nearly everybody in North, Central, and South America had been agreed upon for nearly two hundred years, and that was this: when folks of a town began to turn yellow and hiccup and vomit black, by scores, by hundreds, every day—the only thing to do was to get up and get out of that town. Because the yellow murderer had a way of crawling through walls and slithering along the ground and popping around corners—it could even pass through fires!— it could die and rise from the dead, that yellow murderer; and after everybody (including the very best physicians) had fought it by doing as many contrary things as they could think of as frantically as they could do them—the yellow jack kept on killing, until suddenly it got fed up with killing. In North America that always came with the frosts in the fall. . . .

This was the state of scientific knowledge about yellow fever up to the year 1900. But from between his mutton chop whiskers Carlos Finlay of Habana howled in a scornful wilderness: "You are all wrong—yellow fever is caused by a mosquito!"

2

There was a bad state of affairs in San Cristobal de Habana in Cuba in 1900. The yellow jack had killed thousands more American soldiers than the bullets of the Spaniards had killed. And it wasn't like most diseases, which considerately pounce upon poor dirty people—it had killed more than one-third of the officers of General Leonard Wood's staff, and staff officers—as all soldiers know—are the cleanest of all officers and the best protected. General Wood had thundered orders; Habana had been scrubbed; happy dirty Cubans had been made into unhappy clean Cubans—"No stone had been left unturned"—in vain! There was more yellow fever in Habana than there had been in twenty years!

Cablegrams from Habana to Washington and on June 25th of 1900 Major Walter Reed came to Quemados in Cuba with orders to "give special attention to questions relating to the cause and prevention of yellow fever." It was a big order. Considering who the man Walter Reed was, it was altogether too big an order. Pasteur had tried it! Of course, in certain ways —though you would say they had nothing to do with hunting microbes—Walter Reed had qualifications. He was the best of soldiers; fourteen years and more he had served on the western plains and mountains; he had been a brave angel flying through blizzards to the bedsides of sick settlers—he had shunned the dangers of beer and bottle-pool in the officers' mess and resisted the seductions of alcoholic nights at draw poker. He had a strong moral nature. He was gentle. But it will take a genius to dig out this microbe of the yellow jack, you say—and are geniuses gentle? Just the same, you will see that this job needed particularly a strong moral nature, and then, besides, since 1891 Walter Reed *had* been doing a bit of microbe hunting. He had done some odd jobs of searching at the very best medical school under the most eminent professor of microbe hunting in America—and that professor had known Robert Koch, intimately.

So Walter Reed came to Quemados, and as he went into the yellow fever hospital there, more than enough young American soldiers passed him, going out, on their backs, feet first. . . . There were going to be plenty of cases to work on all right—fatal cases! Dr. James Carroll was with Walter Reed, and he was not what you would call gentle, but you will see in a moment what a soldier-searcher James Carroll was. And Reed found Jesse Lazear waiting for him—Lazear was a European-trained microbe hunter, aged thirty-four, with a wife and two babies in the States, and with doom in his eyes. Finally there was Aristides Agramonte (who was a Cuban)— it was to be his job to cut up the dead bodies, and very well he did that job, though he never became famous because he had had yellow fever already and so ran no risks. These four were the Yellow Fever Commission.

The first thing the Commission did was to fail to find any microbe whatever in the first eighteen cases of yellow fever that they probed into. There were many severe cases in those eighteen; there were four of those eighteen cases who died; there was not one of those eighteen cases that they didn't claw through from stem to gudgeon, so to speak, drawing blood, making cultures, cutting up the dead ones, making endless careful cultures—and not one bacillus did they find. All the time—it was July and the very worst time for yellow fever— the soldiers were coming out of the hospital of Las Animas feet first. The Commission failed absolutely to find any cause, but that failure put them on the right track. That is one of the humors of microbe hunting—the way men make their finds! Theobald Smith found out about those ticks because he had faith in certain farmers; Ronald Ross found out the doings of those gray mosquitoes because Patrick Manson told him to; Grassi discovered the zanzarone carrying malaria because he was patriotic. And now Walter Reed had failed in the very first part—and anybody would say it was the most important part —of his work. What to do? There was nothing to do. And so Reed had time to hear the voice of that Theorizing Old Fool, Dr. Carlos Finlay, of Habana, shouting: "Yellow fever is caused by *a mosquito!*"

The Commission went to call on Dr. Finlay, and that old gentleman—everybody had laughed at him, nobody had listened to him—was very glad to explain his fool theory to the Commission. He told them the ingenious but vague reasons why he thought it was mosquitoes carried yellow fever; he showed them records of those awful experiments, which would convince nobody; he gave them some little black eggs shaped like cigars and said: "Those are the eggs of the criminal!" And Walter Reed took those eggs, and gave them to Lazear, who had been in Italy and knew a thing or two about mosquitoes, and Lazear put the eggs into a warm place to hatch into wigglers, which presently wiggled themselves into extremely pretty mosquitoes, with silver markings on their backs—markings that looked like a lyre. Now Walter Reed had failed, but you have to give him credit for being a sharp-eyed man with plenty of common sense—and then too, as you will see, he was extraordinarily lucky. While he was failing to find bacilli, even in the dreadful cases, with bloodshot eyes and chests yellow as gold, with hiccoughs and with those prophetic retchings—while he was failing, Walter Reed noticed that the nurses who handled those cases, were soiled by those cases, never got yellow fever! They were non-immunes too, those nurses, but they didn't get yellow fever.

"If this disease were caused by bacillus, like cholera, or plague, some of those nurses certainly should get it," argued Walter Reed to his Commission.

Then all kinds of strange tricks of yellow fever struck Walter Reed. He watched cases of the disease pop up most weirdly in Quemados. A man in a house in 102 Real Street came down with it; then it jumped around the corner to 20 General Lee Street, and from there it hopped across the road—and not one of these families had anything to do with each other, hadn't seen each other, even!

"That smells like something carrying the disease through the air to those houses," said Reed. There were various other exceedingly strange things about yellow fever—they had been discovered by an American, Carter. A man came down with yellow fever in a house. For two or three weeks nothing more happened—the man might die, he might have got better and gone away, but at the end of that two weeks, bang! a bunch of other cases broke out in that house. "That two weeks makes it look as if the virus were taking time to grow in some insect," said Reed, to his Commission who thought it was silly, but they were soldiers.

"So we will try Finlay's notion about mosquitoes," said Walter Reed, for all of the just mentioned reasons, but particularly because there was nothing else for the Commission to do.

That was easy to say, but how to go on with it? Everybody knew perfectly well that you cannot give yellow fever to any animal—not even to a monkey or an ape. To make any kind of experiment to prove mosquitoes carry yellow fever you *must* have experimental animals, and that meant nothing more nor less than human animals. But give human beings yellow fever! In some epidemics—there were records of them!—eighty-five men out of a hundred died of it, in some fifty out of every hundred—almost never less than twenty out of every hundred. It would be murder! But that is where the strong moral nature of Walter Reed came to help him. Here was a blameless man, a Christian man, and a man—though he was mild—who was mad to help his fellow men. And if you could *prove* that yellow fever was *only* carried by mosquitoes. . . .

So, on one hot night after a day among dying men at Pinar del Rio, he faced his Commission: "If the members of the Commission take the risk first—if they let themselves be bitten by mosquitoes that have fed on yellow fever cases, that will set an example to American soldiers, and then—" Reed looked at Lazear, and then at James Carroll.

"I am ready to take a bite," said Jesse Lazear, who had a wife and two small children.

"You can count on me, sir," said James Carroll, whose total assets were his searcher's brain, and his miserable pay as an assistant-surgeon in the army. (His liabilities were a wife and five children.)

3

Then Walter Reed (he had been called home to Washington to make a report on work done in the Spanish War) gave elaborate instructions to Carroll and Lazear and Agramonte. They were secret instructions, and savage instructions, when you consider the mild man he was. It was an immoral business—it was a breach of discipline in its way, for Walter Reed then had no permission from the high military authorities to start it. So Reed left for Washington, and Lazear and Carroll set off on the wildest, most daring journey any two microbe hunters had ever taken. Lazear? You could not see the doom in his eyes—the gleam of the searcher outshone it. Carroll? That was a soldier who cared no damn for death or courts-martial—Carroll was a microbe hunter of the great line. . . .

Lazear went down between the rows of beds on which lay men, doomed men with faces yellow as the leaves of autumn, delirious men with bloodshot eyes. He bit those men with his silver-striped she-mosquitoes; carefully he carried these blood-filled beasts back to their glass homes, in which were little saucers of water and little lumps of sugar. Here the she-mosquitoes digested their meal of yellow fever blood, and buzzed a little, and waited for the test.

"We should remember malaria," Reed had told Lazear and Carroll. "In that disease it takes two or three weeks for the mosquito to become dangerous—maybe it's the same here."

But look at the bold face of Jesse Lazear, and tell me if that was a patient man! Not he. Somehow he collected seven volunteers, who so far as I can find have remained nameless, since the test was done in dark secrecy. To these seven men—whom for all I know he may have shanghaied—but first of all to himself, Lazear applied those mosquitoes who a few days before had fed on men who now were dead. . . .

But alas, they all stayed fit as fiddles, and that discouraged Lazear.

But there was James Carroll. For years he had been the right-hand man of Walter Reed. He had come into the army as a buck private and had been a corporal and a sergeant for years—obeying orders was burned into his very bones—and Major Reed had said: "Try mosquitoes!" What is more, what Major Reed thought was right, James Carroll thought was right, too, and Major Reed thought there was something in the notion of that Old Theorizing Fool. But in the army, thoughts are secondary—Major Reed had left them saying: "Try mosquitoes!"

So James Carroll reminded the discouraged Lazear: "I am ready!" He told Lazear to bring out the most dangerous mosquito in his collection—not one that had bitten only a single case, but he must use a mosquito that had bitten many cases —and they must be bad cases—of yellow fever. That mosquito must be as dangerous as possible! On the twenty-seventh of

August, Jesse Lazear picked out what he thought to be his champion mosquito, and this creature, which had fed on four cases of yellow fever, two of them severe ones, settled down on the arm of James Carroll.

That soldier watched her while she felt around with her stinger. . . . What did he think as he watched her swell into a bright balloon with his blood? Nobody knows. But he could think, what everybody knows: "I am forty-six years old, and in yellow fever the older the fewer—get better." He was forty-six years old. He had a wife and five children, but that evening James Carroll wrote to Walter Reed:

"If there is anything in the mosquito theory, I should get a good dose of yellow fever!" He did.

Two days later he felt tired and didn't want to visit patients in the yellow fever ward. Two days after that he was really sick: "I must have malaria!" he cried, and went to the laboratory under his own power, to squint at his own blood under the microscope. But no malaria. That night his eyes were bloodshot, his face a dusky red. The next morning Lazear packed Carroll off to the yellow fever wards, and there he lay, near to death for days and days. . . . There was one minute when he thought his heart had stopped . . . and that, as you will see, was a bad minute for Assistant-Surgeon Carroll.

He always said those were the proudest days of his life. "I was the first case to come down with yellow fever after the experimental bite of a mosquito!" said Carroll.

Then there was that American private soldier they called "X.Y."—these outlaw searchers called him "X.Y.," though he was really William Dean, of Grand Rapids, Michigan. While James Carroll was having his first headaches, they bit this X.Y. with four mosquitoes—the one that nearly killed Carroll, and then three other silver-striped beauties besides, who had fed on six men that were fairly sick, and four men that were very sick with yellow fever and two men that died.

Now everything was fine with the experiments of Quemados. Eight men had been bitten, it is true, and were fit as fiddles—but the last two, James Carroll and X.Y., they were real experimental guinea-pigs, those two, they had both got yellow fever—and James Carroll's heart had nearly stopped, but now they were both getting better, and Carroll was on the heights, writing to Walter Reed, waiting proudly for his chief to come back—to show him the records. Only Jesse Lazear was a little cynical about these two cases, because Lazear was a fine experimenter, a tight one, a man who had to have every condition just so, like a real searcher—and, thought Lazear, "It is too bad seeing the nerve of Carroll and X.Y.—but both of them exposed themselves in dangerous zones once or twice, before they came down. It wasn't an absolutely perfect experiment—it isn't sure that *my* mosquitoes gave them yellow fever!" So Lazear was skeptical, but orders were orders, and every afternoon he went to those rows of beds at Las Animas, in the room with the faint strange smell, and here he turned his test-tubes upside-down on the arms of boys with bloodshot eyes, and let his she-mosquitoes suck their fill. But September 13th was a bad day, it was an unlucky day for Jesse Lazear, for while he was at this silly job of feeding his mosquitoes, a stray mosquito settled down on the back of his hand. "Oh! that's nothing!" he thought. "That wouldn't be the right kind of mosquito anyway!" he muttered, and he let the mosquito drink her fill—though, mind you, she was a stray beast that lived in this ward where men were dying!

That was September 13th.

"On the evening of September 18th . . . Dr. Lazear complained of feeling out of sorts, and had a chill at 8 P.M.," says a hospital record of Las Animas. . . .

"September 19: Twelve o'clock noon," goes on that laconic record, "temperature 102.4 degrees, pulse 112. Eyes injected, face suffused. [That means bloodshot and red] . . . 6 P.M. temperature 103.8 degrees, pulse, 106. Jaundice appeared on the third day. The subsequent history of this case was one of progressive and fatal yellow fever" [and the record softens a little], "the death of our lamented colleague having occurred on the evening of September 25, 1900."

4

Then Reed came back to Cuba, and Carroll met him with enthusiasm, and Walter Reed was sad for Lazear, but very happy about those two successful cases of Carroll and X.Y.— and then, and then (brushing aside tears for Lazear) even in that there was the Hand of God, there was something for Science: "As Dr. Lazear was bitten by a mosquito while present in the wards of a yellow fever hospital," wrote Walter Reed, "one must, at least, admit the possibility of this insect's contamination by a previous bite of a yellow fever patient. This case of accidental infection therefore *cannot fail to be of interest*. . . ."

"Now it is my turn to take the bite!" said Walter Reed, but he was fifty years old, and they persuaded him not to. "But we *must* prove it!" he insisted, so gently, that, hearing his musical voice and looking at his chin that did not stick out like the chin of a he-man, you might think Walter Reed was wavering (after all, here was one man dead out of three).

"But we must prove it," said that soft voice, and Reed went to General Leonard Wood, and told him the exciting events that had happened. Who could be less of a mollycoddle than this Wood? And he gave Walter Reed permission to go as far as he liked. He gave him money to build a camp of seven tents and two little houses—to say nothing of a flagpole—but what was best of all Wood gave him money to buy men, who would get handsomely paid for taking a sure one chance out of five of never having a chance to spend that money! So Walter Reed said: "Thank you, General," and one mile from Quemados they pitched seven tents and raised a flagpole, and flew an American flag and called that place Camp Lazear (three cheers for Lazear!), and you will see what glorious things occurred there.

Now, nothing is more sure than this: that every man of the great line of microbe hunters is different from every other man of them, but every man Jack of them has one thing in common: they are original. They were all original, excepting

Walter Reed—whom you cannot say would be shot for his originality, seeing that this business of mosquitoes and various bugs and ticks carrying diseases was very much in the air in those last ten years of the nineteenth century. It was natural for a man to think of that! But he was by all odds the most moral of the great line of microbe hunters—aside from being a very thorough clean-cut experimenter—and now that Walter Reed's moral nature told him: "You must kill men to save them!" he set out to plan a series of airtight tests—never was there a good man who thought of more hellish and dastardly tests!

And he was exact. Every man about to be bit by a mosquito must stay locked up for days and days and weeks, in that sun-baked Camp Lazear—to keep him away from all danger of accidental contact with yellow fever. There would be no catch in these experiments! And then Walter Reed let it be known, to the American soldiers in Cuba, that there was another war on, a war for the saving of men—were there men who would volunteer? Before the ink was dry on the announcements Private Kissenger of Ohio stepped into his office, and with him came John J. Moran, who wasn't even a soldier—he was a civilian clerk in the office of General Fitzhugh Lee. "You can try it on us, sir!" they told him.

Walter Reed was a thoroughly conscientious man. "But, men, do you realize the danger?" And he told them of the headaches and the hiccups and the black vomit—and he told them of fearful epidemics in which not a man had lived to carry news or tell the horrors. . . .

"We know," said Private Kissenger and John J. Moran of Ohio, "we volunteer solely for the cause of humanity and in the interest of science."

Then Walter Reed told them of the generosity of General Wood. A handsome sum of money they would get—two hundred, maybe three hundred dollars, if the silver-striped she-mosquitoes did things to them that would give them one chance out of five not to spend that money.

"The one condition on which we volunteer, sir," said Private Kissenger and civilian clerk John J. Moran of Ohio, "is that we get no compensation for it."

To the tip of his cap went the hand of Walter Reed (who was a major): "Gentlemen, I salute you!" And that day Kissenger and John J. Moran went into the preparatory quarantine, that would make them first-class, unquestionable guinea-pigs, above suspicion and beyond reproach. On the 5th of December Kissenger furnished nice full meals for five mosquitoes—two of them had bitten fatal cases fifteen days and nineteen days before. Presto! Five days later he had the devil of a backache, two days more and he was turning yellow—it was a perfect case, and in his quarters Walter Reed thanked God, for Kissenger got better! Then great days came to Reed and Carroll and Agramonte—for, if they weren't exactly overrun with young Americans who were ready to throw away their lives in the interest of science—and for humanity still there were ignorant people, just come to Cuba from Spain, who could very well use two hundred dollars. There were five of these mercenary fellows—whom I shall simply have to call "Spanish immigrants," or I could call them Man 1, 2, 3, and 4—just as microbe hunters often mark animals: "Rabbit 1, 2, 3, and 4—" anyway they were bitten, carefully, by mosquitoes who, when you take averages, were much more dangerous than machine gun bullets. They earned their two hundred dollars—for four out of five of them had nice typical (doctors would look scientific and call them beautiful) cases of yellow fever! It was a triumph! It was sure! Not one of these men had been anywhere near yellow fever—like so many mice they had been kept in their screened tents at Quemados. If they hadn't been ignorant immigrants—hardly more intelligent than animals, you might say—they might have been bored, because nothing had happened to them excepting—the stabs of silver-striped she-mosquitoes. . . .

"Rejoice with me, sweetheart," Walter Reed wrote to his wife, "as, aside from the antitoxin of diphtheria and Koch's discovery of the tubercle bacillus, it will be regarded as the most important piece of work, scientifically, during the nineteenth century. . . ."

Walter Reed was so thorough that you can call him original, as original as any of the microbe hunters of the great line—for he was certainly original in his thoroughness. He might have called it a day—you would swear he was tempted to call it a day: eight men had got yellow fever from mosquito bites, and only one—what amazing luck!—had died.

"But can yellow fever be carried in any other way?" asked Reed.

Everybody believed that clothing and bedding and possessions of yellow fever victims were deadly—millions of dollars worth of clothing and bedding had been destroyed; the Surgeon-General believed it; every eminent physician in America, North, South and Central (excepting that old fool Finlay) believed it. "But can it?" asked Reed, and while he was being so joyfully successful with Kissenger and Spaniards 1, 2, 3, and 4, carpenters came, and built two ugly little houses in Camp Lazear. House No. 1 was the nastier of these two little houses. It was fourteen feet by twenty, it had two doors cleverly arranged one back of the other so no mosquitoes could get into it, it had two windows looking south—they were on the same side as the door, so no draft could blow through that little house. Then it was furnished with a nice stove, to keep the temperature well above ninety, and there were tubs of water in the house—to keep the air as chokey as the hold of a ship in the tropics. So you see it was an uninhabitable little house—under the best of conditions—but now, on the thirtieth of November in 1900, sweating soldiers carried several tightly nailed suspicious-looking boxes, that came from the yellow fever wards of Las Animas—to make this house altogether cursed. . . .

That night, of the thirtieth of November, Walter Reed and James Carroll were the witnesses of a miracle of bravery, for into this House No. 1 walked a young American doctor named

Cooke, and two American soldiers, whose names—where are their monuments?—were Folk and Jernegan.

Those three men opened the tightly nailed, suspicious-looking boxes. They opened those boxes inside that house, in air already too sticky for proper breathing.

Phew! There were cursings, there were holdings of noses.

But they went on opening those boxes, and out of them Cooke and Folk and Jernegan took pillows, soiled with the black vomit of men dead of yellow fever; out of them they took sheets and blankets, dirty with the discharges of dying men past helping themselves. They beat those pillows and shook those sheets and blankets—"you must see the yellow fever poison is well spread around that room!" Walter Reed had told them. Then Cooke and Folk and Jernegan made up their little army cots with those pillows and blankets and sheets. They undressed. They lay down on those filthy beds. They tried to sleep—in that room fouler than the dankest of medieval dungeons. . . . And Walter Reed and James Carroll guarded that little house, so tenderly, to see no mosquito got into it, and Folk and Cooke and Jernegan had the very best of food, you may be sure. . . .

Night after night those three lay in that house, wondering perhaps about the welfare of the souls of their predecessors in those sheets and blankets. They lay there, wondering whether anything else besides mosquitoes (though mosquitoes hadn't even been proved to carry it then!) carried yellow fever. . . . Then Walter Reed, who was a moral man and a thorough man, and James Carroll, who was a grim man, came to make their test a little more thorough. More boxes came to them from Las Animas—and when Cooke and Jernegan and Folk unpacked them, they had to rush out of their little house, it was so dreadful.

But they went back in, and they went to sleep. . . .

For twenty nights—where are their monuments?—these three men stayed there, and then they were quarantined in a nice airy tent, to wait for their attack of yellow fever. But they gained weight. They felt fit as fiddles. They made vast jokes about their dirty house and their perilous sheets and blankets. They were happy as so many schoolboys when they heard Kissenger and those Spaniards (1, 2, 3, and 4) had really got the yellow jack after the mosquito bites. What a marvelous proof, you will say, but what a dastardly experiment—but for the insanely scientific Walter Reed that most dastardly experiment was not marvelous enough! Three more American boys went in there, and for twenty nights slept in new unspeakable sheets and blankets—with this little refinement of the experiment: they slept in the very pajamas in which yellow fever victims had died. And then for twenty more nights three other American lads went into House No. 1, and slept that way— with this additional little refinement of the experiment: they slept on pillows covered with towels soaked with the blood of men whom the yellow jack had killed.

But they all stayed fit as fiddles! Not a soul of these nine men had so much as a touch of yellow fever! How wonderful is science, thought Walter Reed. "So," he wrote, "the bubble of the belief that clothing can transmit yellow fever was pricked by the first touch of human experimentation." Walter Reed was right. It is true, science is wonderful. But science is cruel, microbe hunting can be heartless, and that relentless devil that was the experimenter in Walter Reed kept asking: "But is your experiment really sound?" None of those men who slept in House No. 1 got yellow fever, that is true—but how do you know they were *susceptible* to yellow fever? Maybe they were naturally immune! Then Reed and Carroll, who had already asked as much of Folk and Jernegan as any captain has ever asked of any soldier—so it was that Reed and Carroll now shot virulent yellow fever blood under the skin of Jernegan, so it was they bit Folk with mosquitoes who had fed on fatal cases of yellow fever. They both came down with wracking pains and flushed faces and bloodshot eyes. They both came through their Valley of the Shadow. "Thank God," murmured Reed—but especially Walter Reed thanked God he had proved those two boys were not immune during those twenty hot stinking nights in House No. 1.

For these deeds Warren Gladsden Jernegan and Levi E. Folk were generously rewarded with a purse of three hundred dollars—which in those days was a lot of money.

5

While these tests were going on John J. Moran, that civilian clerk from Ohio, whom Walter Reed had paid the honor of a salute, was a very disappointed man. He had absolutely refused to be paid; he had volunteered in "the interest of science and for the cause of humanity," he had been bitten by those silver-striped Stegomyia mosquitoes (the bug experts just then thought this was the proper name for that mosquito)—he had been stabbed several times by several choice poisonous ones, but he hadn't come down with yellow fever, alas, he stayed fit as a fiddle. What to do with John J. Moran?

"I have it!" said Walter Reed. "This to do with John J. Moran!"

So there was built, close by that detestable little House No. 1, another little house, called House No. 2. That was a comfortable house! It had windows on the side opposite to its door, so that a fine trade wind played through it. It was cool. It had a nice clean cot in it, with steam-disinfected bedding. It would have been an excellent house for a consumptive to get better in. It was a thoroughly sanitary little house. Half way across the inside of it was a screen, from top to bottom, a fine-meshed screen that the tiniest mosquito found it impossible to fly through. At 12 o'clock noon on the twenty-first of December in 1900, this John J. Moran (who was a hog for these tests) "clad only in a nightshirt and fresh from a bath" walked into this healthy little house. Five minutes before Reed and Carroll had opened a glass jar in that room, and out of that jar flew fifteen she-mosquitoes, thirsty for blood, whining for a meal of blood, and each and every one of those fifteen mosquitoes, had fed, on various days before—on the blood of yellow-faced boys in the hospital of Las Animas.

Clad only in a nightshirt and fresh from a bath, Moran—who knows of him now?—walked into the healthy little room and lay down on his clean cot. In a minute that damned buzzing started round his head, in two minutes he was bitten, in the thirty minutes he lay there he was stabbed seven times—without even the satisfaction of smashing those mosquitoes. You remember Mr. Sola, whom Grassi tortured—he probably had his worried moments—but all Mr. Sola had to look forward to was a little attack of malaria and a good dose of curative quinine to get him out of it. But Moran? But John J. Moran was a hog for such tests! He was back there at four-thirty the same afternoon, to be bitten again, and once more the next day—to satisfy the rest of the hungry she-mosquitoes who hadn't found him the first day. In the other room of this house, with only a fine meshed but perfect wire screen between them and Moran—and the mosquitoes—lay two other boys, and those two boys slept in that house safely for eighteen nights.

But Moran?

On Christmas morning of 1900, there was a fine present waiting for him—in his head, how that thumped—in his eyes, how red they were and how the light hurt them—in his bones, how tired they were! A nasty knock those mosquitoes had hit him and he came within a hair of dying but (thank God! murmured Walter Reed) he was saved, this Moran, to live the rest of his life in an obscurity he didn't deserve. So Moran had his wish—in the interest of science, and for humanity! So he, with Folk and Jernegan and Cooke and all those others proved that the dirty pest hole of a house (with no mosquitoes) was safe; and that the clean house (but with mosquitoes) was dangerous, so dangerous! So at last Walter Reed had every answer to his diabolical questions, and he wrote, in that old-fashioned prose of his: "The essential factor in the infection of a building with yellow fever is the presence therein of mosquitoes that have bitten cases of yellow fever."

It was so simple. It was true. That was all. That was that. And Walter Reed wrote to his wife:

"The prayer that has been mine for twenty years, that I might be permitted in some way or at some time to do something to alleviate human suffering has been granted! A thousand Happy New Years. . . . Hark, there go the twenty-four buglers in concert, all sounding taps for the old year!"

They were sounding taps, were those buglers, for the searcher that was Jesse Lazear, and for the scourge of yellow fever that could now be wiped from the earth. They were blowing their bugles, those musicians, to celebrate—as you will see—the fate that waited for that little commission after a too short hour of triumph. . . .

6

Then the world came to Habana, and there was acclaim for Walter Reed, and the customary solemn discussions and doubts and arguments of the learned men who came. William Crawford Gorgas (who was another blameless man!) grooming himself for the immortality of Panama, went into the gutters and cesspools and cisterns of Habana, making horrid war on the Stegomyia mosquitoes, and in ninety days, Habana had not a single case of yellow jack—she was free for the first time in two hundred years. It was magical! But still there came learned doctors, and solemn bearded physicians, from Europe and America, asking this, questioning that—and one morning fifteen of these skeptics were in the mosquito room of the laboratory—oh! they were from Missouri! "These are remarkable experiments, but the results should be weighed and considered with reserve . . . et cetera!" Then the gauze lid came off a jar of she-mosquitoes (of course it was by accident) and into the room, with wicked lustful eyes on those learned scientists the Stegomyia buzzed. Alas for skepticism! Away went all doubts! From the room rushed the eminent servants of knowledge! Down went the screen door with a crash—such was the vehemence of their conviction that Walter Reed was right. (Though it happened that this particular jar of mosquitoes was not contaminated.)

Then William Crawford Gorgas and John Guitéras—he was a great Cuban authority on yellow jack—they were convinced too by those experiments at Camp Lazear, and they were full of excellent plans to put those experiments in practice—fine plans, but rash plans, alas. "It is remarkable," said Gorgas and Guitéras, "that these experimental cases at Camp Lazear didn't die—they had typical yellow fever, but they got better, maybe because Reed put them to bed so quickly." Then they proceeded to play with fire. "We will give newly arrived non-immune immigrants yellow fever—a smart attack of it, but a *safe* attack of it." They planned this, when it really was so easy to wipe out yellow fever simply by warring on the Stegomyia, which does not breed in secret places, which is a very domestic mosquito! "And at the same time we can confirm Reed's results," thought Gorgas and Guitéras.

The immigrants (of course they were very ignorant people) came; the immigrants listened and were told it was safe; seven immigrants and a bold young American nurse were bitten by the poisoned Stegomyia. And of these eight, two immigrants and the bold young American nurse went out from the hospital, safe from another attack of yellow fever, safe from all the worries of the world. . . . They went out, feet first—to slow music. What a fine searcher was Walter Reed—but what amazing luck he had, in those experiments at Camp Lazear. . . .

There was panic in Habana, and mutterings of the mob—and who can blame that mob, for human life is sacred. But there was Assistant Surgeon James Carroll, unsentimental as an embalmer and before all else a soldier,—he had just then come back to Habana to settle certain little academic questions. "We can wipe out yellow fever now, we have proved just how it gets from man to man—*but what is it causes yellow fever?*" This is what Reed and Carroll asked each other, and everybody must admit that it was a purely academic question, and I ask you: was it worth a human life (even of a Spanish immigrant) to find the answer? Myself I cannot answer yes or no. But Reed and Carroll answered yes! Starting out as soldiers obeying orders, as humanitarians risking their hides to save the lives of men, they had been bitten by the virus of the search for truth, cold truth—they were enchanted with the glory that comes from the discovery of unknown things. . . .

They were sure there was no visible bacillus, nor any kind of microbe that could be seen through the strongest microscope to cause it—they had looked in the livers of men and the lights of mosquitoes for such a germ, in vain. But there were other possibilities—magical possibilities, of a new kind of germ that might be the cause of yellow fever, an ultra-microbe, too immensely small for the strongest lens to uncover, revealing its existence only by the murdering of men with its unseen mysterious poison. That might be the nature of the germ of yellow fever. Old Friedrich Loeffler—he of the mustaches—had found such little life making calves sick with foot-and-mouth disease. And now if Reed and Carroll could show the microbe of yellow fever belonged to this sub-microscopic world too!

Walter Reed was busy, so he sent James Carroll to Habana to see, and here you find James Carroll, intensely annoyed because those experimental cases of Guitéras had died. Guitéras—do you blame him?—was in a funk. No, Carroll mightn't draw blood from yellow fever patients. Indeed not, Carroll mightn't even bite them with mosquitoes. What was most silly, Dr. Guitéras would rather not have Dr. Carroll make post-mortems on the dead cases—it might enrage the population of Habana. "You can imagine my disappointment!" wrote Carroll to Walter Reed, with indignant remarks about the frivolous fears of ignorant populations. But did those deaths stop him? Not Carroll!

By some unexplained sorceries he got hold of some good poisonous yellow fever blood, and filtered it through a porcelain filter that was so fine no visible microbe could get through it. The stuff that came through that filter Carroll shot under the skin of three non-immunes (history doesn't tell how he induced them to stand for it)—and presto! two of them got yellow fever. Hurrah! Yellow fever was like foot-and-mouth disease then. Its cause was a germ maybe too little to see, a microbe that could sneak through fine-grained porcelain.*

Reed wrote to stop him: those deaths were too much—but Carroll simply *must* get some contaminated mosquitoes, and by some bold devilry he did get them, and heigho for this final most horrible experiment!

"In my own case," said Carroll, "produced by the bite of a single mosquito, a fatal result was looked for during several days. I became so firmly convinced that the severity of the attack depended upon the susceptibility of an individual rather than on the number of bites he had got, that on October 9, 1901, at Habana, *I purposely applied to a non-immune eight mosquitoes (all I had) that had been contaminated eighteen days before. The attack that followed was a mild one,*" ended Carroll, triumphantly. But what if that patient had died—as God knows he might have?

Such was the strangest of that strange crew, and looking back on this his boldness, in despite of his fanatic prying into dangerous mysteries, my hat is off to this bald-headed bespectacled ex-lumberjack searcher. He himself was the first to be hit, it was Carroll gave the example to those American soldiers, to that civilian clerk, and to those Spanish immigrants —1, 2, 3, and 4—and to all the rest of the unknown numbers of them. And do you remember, in the middle of his attack of yellow fever, that moment when his heart seemed to stop? In 1907, six years after, Carroll's heart stopped for good. . . .

7

And in 1902, five years before that, Walter Reed, in the prime of his life, but tired, so tired, died—just as the applause of

---

* A spiral-shaped microbe has recently been brought forward as the cause of yellow fever, but this discovery has not yet been confirmed.

nations grew thunderous—of appendicitis. "I am leaving my wife and daughter so little . . ." said Walter Reed to his friend Kean, just before the ether cone went down over his face. "So little . . ." he mumbled as the ether let him down into his last dreams. But let us be proud of our nation, and proud of our Congress—for they voted Mrs. Emilie Laurence Reed, wife of the man who has saved the world no one knows what millions of dollars—let us say nothing of lives—they voted her a handsome pension, of fifteen hundred dollars a year! And the same for the widow of Lazear, and the same for the widow of James Carroll—and surely that was handsome for them, because, as one committee of senators quaintly said: "They can still help themselves."

But what of Private Kissenger, of Ohio, who stood that test, in the interest of science—and for humanity? He didn't die from yellow fever. And they prevailed upon him, at last, to accept one hundred and fifteen dollars and a gold watch, which was presented to him in the presence of the officers and men of Columbia barracks. He didn't die—but what was worse, as the yellow fever germs went out of him, a paralysis crept into him—now he sits, counting the hours on his gold watch. But what luck! At the last account he had a good wife to support him by taking in washing.

And what of the others? Time is too short to deal with those others—and besides I do not know what has become of them. So it is that this strange crew has made rendezvous, each one, with his special and particular fate—this strange crew who put the capstone on that most marvelous ten years of the microbe hunters, that crew who worked together so that now, in 1926, there is hardly enough of the poison of yellow fever left in the world to put on the points of six pins. . . .

So it is that the good death-fighter, David Bruce, should eat his words: "It is impossible, at present, to experiment with human beings."

12

Paul Ehrlich

.The Magic Bullet

1

Two hundred and fifty years ago, Antony Leeuwenhoek, who was a matter-of-fact man, looked through a magic eye, saw microbes, and so began this history. He would certainly have snorted a contemptuous Dutch sort of snort at anybody who called his microscope a magic eye.

Now Paul Ehrlich—who brings this history to the happy end necessary to all serious histories—was a gay man. He smoked twenty-five cigars a day; he was fond of drinking a seidel of beer (publicly) with his old laboratory servant and many seidels of beer with German, English and American colleagues; a modern man, there was still something medieval about him for he said: "We must learn to shoot microbes with magic bullets." He was laughed at for saying that, and his enemies cartooned him under the name "Doktor Phantasus."

But he did make a magic bullet! Alchemist that he was, he did something more outlandish than that, for he changed a drug that is the favorite poison of murderers into a saver of the lives of men. Out of arsenic he concocted a deliverer from the scourge of that pale corkscrew microbe whose attack is the reward of sin, whose bit is the cause of syphilis, the ill of the loathsome name. Paul Ehrlich had a most weird and wrong-headed and unscientific imagination: that helped him to make microbe hunters turn another corner, though alas, there have been few of them who have known what to do when they got around that corner, which is why this history has to stop with Paul Ehrlich.

Of course, it is sure as the sun following the dawn of to-morrow, that the high deeds of the microbe hunters have not come to an end; there will be others to fashion magic bullets. And they will be waggish men and original, like Paul Ehrlich, for it is not from a mere combination of incessant work and magnificent laboratories that such marvelous cures are to be got. . . . To-day? Well, to-day there are no microbe hunters who look you solemnly in the eye and tell you that two plus two makes five. Paul Ehrlich was that kind of a man. Born in March of 1854 in Silesia in Germany, he went to the gym-nasium at Breslau, and his teacher of literature ordered him to write an essay, subject: "Life is a Dream."

"Life rests on normal oxidations," wrote that bright young Jew, Paul Ehrlich. "Dreams are an activity of the brain and the activities of the brain are only oxidations . . . dreams are a sort of phosphorescence of the brain!"

He got a bad mark for such smartness, but then he was always getting bad marks. Out of the gymnasium, he went to a medical school, or rather, to three or four medical schools —Ehrlich was that kind of a medical student. It was the opin-ion of the distinguished medical faculties of Breslau and Strasbourg and Freiburg and Leipsic that he was no ordinary student. It was also their opinion he was an abominably bad student, which meant that Paul Ehrlich refused to memorize the ten thousand and fifty long words supposed to be needed for the cure of sick patients. He was a revolutionist, he was part of the revolt led by that chemist, Louis Pasteur, and the country doctor, Robert Koch. His professors told Paul Ehrlich to cut up dead bodies and learn the parts of dead bodies; in-stead he cut up one part of a dead body into very thin slices and set to work to paint these slices with an amazing variety of pretty-colored aniline dyes, bought, borrowed, stolen from under his demonstrator's nose.

He hadn't a notion of why he liked to do that—though there is no doubt that to the end of his days this man's chief joy (aside from wild scientific discussions over the beer tables) was in looking at brilliant colors, and making them.

"Ho, Paul Ehrlich—what are you doing there?" asked one of his professors, Waldeyer.

"Ja, Herr Professor, I am *trying* with different dyes!"

He hated classical training, he called himself a modern, but he had a fine knowledge of Latin, and with this Latin he used to coin his battle cries. For he worked by means of battle cries and slogans rather than logic. "*Corpora non agunt nisi fixata!*" he would shout, pounding the table till the dishes danced— "Bodies do not act unless fixed!" That phrase heartened him through thirty years of failure. "You see! You understand! You know!" he would say, waving his horn-rimmed spectacles in your face, and if you took him seriously you might think that Latin rigmarole (and not his searcher's brain) carried him to his final triumph. And in a way there is no doubt it did!

Paul Ehrlich was ten years younger than Robert Koch; he was in Cohnheim's laboratory on that day of Koch's first demonstration of the anthrax microbe; he was atheistical, so he needed some human god and that god was Robert Koch. Painting a sick liver Ehrlich had seen the tubercle germ before ever Koch laid eyes on it. Ignorant, lacking Koch's clear intelligence, he supposed those little colored rods were crystals. But when he sat that evening in the room in Berlin in March, 1882, and listened to Koch's proof of the discovery of the cause of consumption, he saw the light: "It was the most gripping experience of my scientific life," said Paul Ehrlich, long afterwards. So he went to Koch. He must hunt microbes too! He showed Robert Koch an ingenious way to stain that tubercle microbe—that trick is used, hardly changed, to this day. He would hunt microbes! And in the enthusiastic way he had he proceeded to get consumption germs all over himself: so he caught consumption and had to go to Egypt.

2

Ehrlich was thirty-four years old then, and if he had died in Egypt, he would certainly have been forgotten, or been spoken of as a color-loving, gay, visionary failure. He had the energy of a dynamo; he had believed you could treat sick people and hunt microbes at the same time; he had been head physician in a famous clinic in Berlin, but he was a very raw-nerved man and was fidgety under the cries of sufferers past helping and the deaths of patients who could not be cured. To cure them! Not by guess or by the bedside manner or by the laying on of hands or by waiting for Nature to do it—but how to *cure* them! These thoughts made him a bad doctor, because doctors should be sympathetic but not desperate about ills over which they are powerless. Then, too, Paul Ehrlich was a disgusting doctor because his brain was in the grip of dreams: he looked at the bodies of his patients: he seemed to see through their skins: his eyes became super-microscopes that saw the quivering stuff of the cells of these bodies as nothing more than complicated chemical formulas. Why of course! Living human stuff was only a business of benzene rings and side-chains, just like his dyes! So Paul Ehrlich (caring nothing for the latest physiological theories) invented a weird old-fashioned life-chemistry of his own; so Paul Ehrlich was anything but a Great Healer; so he would have been a failure— But he didn't die!

"I will stain live animals!" he cried. "The chemistry of animals is like the chemistry of my dyes—staining them while they are still alive—that will tell me all about them!" So he took his favorite dye, which was methylene blue, and shot a little of it into the ear vein of a rabbit. He watched the color flow through the blood and body of the beast and mysteriously pick out and paint the living endings of its nerves blue—but no other part of it! How strange! He forgot all about his fundamental science for a moment. "Maybe methylene blue will kill pain then," he muttered, and he straightway injected this blue stuff into groaning patients, and maybe they were eased a little, but there were difficulties, of a more or less entertaining nature, which maybe frightened the patients—who can blame them?

He failed to invent a good pain-killer, but from this strange business of methylene blue pouncing on just one tissue out of all the hundred different kinds of stuff that living things are made of, Paul Ehrlich invented a fantastic idea which led him at last to his magic bullet.

"Here is a dye," he dreamed, "to stain only one tissue out of all the tissues of an animal's body—there must be one to hit *no* tissue of men, but to stain and kill the microbes that attack men." For fifteen years and more he dreamed that, before ever he had a chance to try it. . . .

In 1890 Ehrlich came back from Egypt; he had not died from tuberculosis; Robert Koch shot his terrible cure for consumption into him, still he did not die from tuberculosis—and presently he went to work in the Institute of Robert Koch in Berlin, in those momentous days when Behring was massacring guinea-pigs to save babies from diphtheria and the Japanese Kitasato was doing miraculous things to mice with lockjaw. Ehrlich was the life of that grave place! Koch would come into his pupil's crammed and topsy-turvy laboratory, that gleamed and shimmered with rows of bottles of dyes Ehrlich had no time to use—for you may be sure Koch was Tsar in that house and thought Ehrlich's dreams of magic bullets were nonsense. Robert Koch would come in and say:

"Ja, my dear Ehrlich, what do your experiments tell us to-day?"

Then would come a geyser of excited explanations from Paul Ehrlich, who was prying then into the way mice may become immune to those poisons of the beans called the castor and the jequirity:

"You see, I can measure exactly—it is always the same!— the amount of poison to kill in forty-eight hours a mouse weighing ten grams. . . . You know, I can now plot a curve of the way the immunity of my mice increases—it is as exact as experiments in the science of physics. . . . You understand, I have found how it is this poison kills my mice; it clots his blood corpuscles inside his arteries! That is the whole explanation of it . . ." and Paul Ehrlich waved test-tubes filled with brick-red clotted clumps of mouse blood at his famous chief, proving to him that the amount of poison to clot that blood was just the amount that would kill the mouse that the blood came from. Torrents of figures and experiment Paul Ehrlich poured over Robert Koch—

"But wait a moment, my dear Ehrlich! I can't follow you —please explain more clearly!"

"Certainly, Herr Doktor! That I can do right off!" Never for a moment does Ehrlich stop talking, but grabs a piece of chalk, gets down on his knees, and scrawls huge diagrams of his ideas over the laboratory floor—"Now, do you see, is that clear?"

There was no dignity about Paul Ehrlich! Neither about his attitudes, for he would draw pictures of his theories anywhere, with no more sense of propriety than an annoying little boy, on his cuffs and the bottoms of shoes, on his own shirt front to the distress of his wife, and on the shirt fronts of his colleagues if they did not dodge fast enough. Nor could you properly say Paul Ehrlich was dignified about his thoughts, because, twenty-four hours a day he was having the most outrageous thoughts of why we are immune or how to measure immunity or how a dye could be turned into a magic bullet. He left a trail of fantastic pictures of those thoughts behind him everywhere!

Just the same he was the most exact of men in his experiments. He was the first to cry out against the messy ways of microbe hunters, who searched for truth by pouring a little of this into some of that, and in that laboratory of Robert Koch he murdered fifty white mice where one was killed before, trying to dig up simple laws, to be expressed in numbers, that he felt lay beneath the enigmas of immunity and life and death.

And that exactness, though it did nothing to answer those riddles, helped him at last to make the magic bullet.

3

Such was the gayety of Paul Ehrlich, and such his modesty—for he was always making straight-faced jokes at his own ridiculousness—that he easily won friends, and he was a crafty man too and saw to it that certain of these friends were men in high places. Presently, in 1896, he was director of a laboratory of his own; it was called the Royal Prussian Institute for Serum Testing. It was at Steglitz, near Berlin, and it had one little room that had been a bakery and another little room that had been a stable. "It is because we are not exact that we fail!" cried Ehrlich, remembering the bubble of the vaccines of Pasteur which had burst, and the balloon of the serums of Behring which had been pricked. "There must be mathematical laws to govern the doings of these poisons and vaccines and antitoxins!" he insisted, so this man with the erratic imagination walked up and down in those two dark rooms, smoking, explaining, expostulating, and measuring as accurately as God would let him with drops of poison broth and calibrated tubes of healing serum.

But laws? He would make an experiment. It would turn out beautifully. "You see! here is the reason of it!" he would say, and draw a queer picture of what a toxin must look like and what the chemistry of a body cell must look like, but as he went on working, as regiments of guinea-pigs marched to their doom, Paul Ehrlich found more exceptions to his simple theories than agreements with them. That didn't bother him, for, such was his imagination, that he invented new little supporting laws to take care of the exceptions, he drew stranger and stranger pictures, until his famous "Side-Chain" theory of immunity became a crazy puzzle, which could explain hardly anything, which could predict nothing at all. To his dying day Paul Ehrlich believed in his silly side-chain theory of immunity; from all parts of the world critics knocked that theory to smithereens—but he never gave it up; when he couldn't find experiments to destroy his critics he argued at them with enormous hair-splittings like Duns Scotus and St. Thomas Aquinas. When he was beaten in these arguments at medical congresses it was his custom to curse—gayly—at his antagonist all the way home. "You see, my dear colleague!" he would cry, "that man is a SHAMELESS BADGER!" Every few minutes, at the top of his voice he yelled this, defying the indignant conductor to put him off the train.

So, in 1899, when he was forty-five, if he had died then, Ehrlich would certainly still have been called a failure. His efforts to find laws for serums had resulted in a collection of fantastic pictures that nobody took very seriously, they certainly had done nothing to turn feebly curative serums into powerful ones—what to do? First, this to do, thought Ehrlich, and he pulled his wires and cajoled his influential friends, and presently the indispensable and estimable Mr. Kadereit, his chief cook and bottle-washer, was dismounting that laboratory at Steglitz—they were moving to Frankfort-on-the-Main, away from the vast medical schools and scientific buzzings of Berlin. What to do? Well, Frankfort was near those factories where the master-chemists turned out their endless bouquets of pretty colors—what could be more important for Paul Ehrlich? Then there were rich Jews in Frankfort, and these rich Jews were famous for their public spirit, and money—*Geld*, that was one of his four big "G's," along with *Geduld*—patience, *Geschick*—cleverness and *Glück*—luck, which Ehrlich always said were needed to find the magic bullet. So Paul Ehrlich came to Frankfort-on-the-Main, or rather, "WE came to Frankfort-on-the-Main," said the valuable Mr. Kadereit, who had the very devil of a time moving all of those dyes and that litter of be-penciled and dog-eared chemical journals.

Reading this history, you might think there was only one good kind of microbe hunter: the kind of searcher who stood on his own absolutely, who paid little attention to the work of other microbe hunters, who read nature and not books. But Paul Ehrlich was not that kind of man! He rarely observed nature, unless it was the pet toad in his garden, whose activities helped Ehrlich to prophesy the weather—it was Mr. Kadereit's first duty to bring plenty of flies to that toad. . . . No, Paul Ehrlich got his ideas out of books.

He lived among scientific books and subscribed to every chemical journal in every language he could read, and in several he couldn't read. Books littered his laboratory so that when visitors came and Ehrlich said: "I beg you, be seated!" there was no place for them to sit at all. Journals stuck out of the pockets of his overcoat—when he remembered to wear one—and the maid, bringing his coffee in the morning, fell over ever-growing mountains of books in his bedroom. Books, with the help of those expensive cigars, kept Paul Ehrlich poor. Mice built nests in the vast piles of books on the old sofa in his office. When he wasn't painting the insides of his animals and the outside of himself with his dyes, he was peering in these books. And what was important inside of those books, was in the brain of Paul Ehrlich, ripening, changing itself into those outlandish ideas of his, waiting to be used. That was where Paul Ehrlich got his ideas—you would never accuse him of stealing the ideas of others!—and queer things happened to those ideas of others when they stewed in Ehrlich's brain.

So now, in 1901, at the beginning of his eight-year search for the magic bullet he read of the researches of Alphonse Laveran. Laveran was the man, you remember, who discovered the malaria microbe, and very lately Laveran had taken to fussing with trypanosomes. He had shot those finned devils, which do evil things to the hind-quarters of horses and give them a disease called the mal de Caderas, into mice. Laveran had watched those trypanosomes kill those mice, one hundred times out of one hundred. Then Laveran had injected arsenic under the skins of some of those suffering mice. That had helped them a little, and killed many of the trypanosomes that gnawed at them, but not one of these mice ever got really better; one hundred out of one hundred died and that was as far as Alphonse Laveran ever got.

But reading this was enough to get Ehrlich started. "Ho!

here is an excellent microbe to work with! It is large and easy to see. It is easy to grow in mice. It kills them with the most beautiful regularity! It *always* kills mice! What could be a better microbe than this trypanosome to use to try to find a magic bullet to cure? Because, if I could find a dye that would save, completely save, just one mouse!"

4

So Paul Ehrlich, in 1902, set out on his hunt. He got out his entire array of gleaming and glittering and shimmering dyes. "Splen-did!" he cried as he squatted before cupboards holding an astounding mosaic of sloppy bottles. He provided himself with plenty of the healthiest mice. He got himself a most earnest and diligent Japanese doctor, Shiga, to do the patient job of watching those mice, of snipping a bit off the ends of their tails to get a drop of blood to look for the trypanosomes, of snipping another bit off the ends of the same tails to get a drop of blood to inject into the next mouse—to do the job, in short, that it takes the industry and patience of a Japanese to do. The evil trypanosomes of the mal de Caderas came in a doomed guinea-pig from the Pasteur Institute in Paris; into the first mouse they went, and the hunt was on.

They tried nearly five hundred dyes! What a completely unscientific hunter Paul Ehrlich was! It was like the first boatman hunting for the right kind of wood from which to make stout oars; it was like primitive blacksmiths clawing among metals for the best stuff from which to forge swords. It was, in short, the oldest of all the ways of man to get knowledge. It was the method of Trial and Sweating! Ehrlich tried; Shiga sweat. Their mice turned blue from this dye and yellow from that one, but the beastly finned trypanosomes of the mal de Caderas swarmed gayly in their veins, and killed those mice, one hundred out of every hundred!

That man Ehrlich smoked more of his imported cigars, even at night in bed he would awake to smoke them; he drank more mineral water; he read in more books, and he threw books at the head of poor Kadereit—who heaven knows could not be blamed for not knowing what dye would kill trypanosomes. He said Latin phrases; he propounded amazing theories of what these dyes ought to do. Never had any searcher coined so many utterly wrong theories. But then, in 1903, came a day when one of these wrong explanations came to help him.

Ehrlich was testing the pretty-colored but complicated benzopurpurin dyes on dying mice, but the mice were dying, with sickening regularity, from the mal de Caderas. Paul Ehrlich wrinkled his forehead—already it was like a corrugated iron roof from the perplexities and failures of twenty years—and he told Shiga:

"These dyes do not spread enough through the mouse's body! Maybe, my dear Shiga, if we change it a little—maybe, let us say, if we added sulfo-groups to this dye, it would dissolve better in the blood of the mouse!" Paul Ehrlich wrinkled his brow.

Now, while Paul Ehrlich's head was an encyclopedia of chemical knowledge, his hands were not the hands of an expert chemist. He hated complicated apparatus as much as he loved complicated theories. He didn't know how to manage apparatus. He was only a chemical dabbler making endless fussy little starts with test-tubes, dumping in first this and then that to change the color of a dye, rushing out of his room to show the first person he met the result, waving the test-tube at him, shouting: "You understand? This is splen-did!" But as for delicate syntheses, those subtle buildings-up and changings of dyes, that was work for the master chemists. "But we must change this dye a little—then it will work!" he cried. Now Paul Ehrlich was a gay man and a most charming one, and presently back from the dye factory near by came that benzopurpurin color, with the sulfo-groups properly stuck onto it, "changed a little."

Under the skin of two white mice Shiga shot the evil trypanosomes of the mal de Caderas. A day passes. Two days go by. The eyes of those mice begin to stick shut with the mucilage of doom, their hair stands up straight with their dread of destruction—one day more and it will be all over with both of those mice. . . . But wait! Under the skin of one of those two mice Shiga sends a shot of that red dye—changed a little. Ehrlich watches, paces, mutters, gesticulates, shoots his cuffs. In a few minutes the ears of that mouse turn red, the whites of his nearly shut eyes turn pinker than the pink of his albino pupils. That day is a day of fate for Paul Ehrlich, it is the day the god of chance is good, for, like snows before the sun of April, so those fell trypanosomes melt out of the blood of that mouse!

Away they go, shot down by the magic bullet, till the last one has perished. And the mouse? His eyes open. He snouts in the shavings in the bottom of his cage and sniffs at the pitiful little body of his dead companion, the untreated one.

He is the first one of all mice to fail to die from the attack of the trypanosome.

Paul Ehrlich, by the grace of persistence, chance, God, and a dye called "Trypan Red" (its real chemical name would stretch across this page!) has saved him! How that encouraged this already too courageous man! "I have a dye to cure a mouse—I shall find one to save a million men," so dreamed that confident German Jew.

But not at once, alas and alas. With gruesome diligence Shiga shot in that trypan red, and some mice got better but others got worse. One, seeming to be cured, would frisk about its cage, and then, after sixty days (!) would turn up seedy in the morning. Snip! went an end off its tail, and the skillful Shiga would call Paul Ehrlich to see its blood matted with a writhing swarm of the fell trypanosomes of the mal de Caderas. Terrible beasts are trypanosomes, sly, tough, as all despicable microbes are tough. And among the tough lot of them there are super-hardy ones. These beasts, when a Jew and a Japanese come along to have at them with a bright-colored dye, lap up that dye. They like it! Or they retreat discreetly to some out-of-the-way place in a mouse's carcass. There they wait their time to multiply in swarms. . . .

So, for his first little success, Paul Ehrlich paid with a thousand disappointments. The trypanosome of David Bruce's nagana and the deadly trypanosome of human sleeping sickness laughed at that trypan red! They absolutely refused to be touched by it! Then, what worked so beautifully with mice, failed completely when they came to try it on white rats and guinea-pigs and dogs. It was a grinding work, to be tackled only by such an impatient persistent man as Ehrlich, for had he not saved one mouse?—What waste! He used thousands of animals! I used to think, in the arrogance of my faith in science: "What waste!" But no. Or call it waste if you like, remembering that nature gets her most sublime results—so often—by being lavishly wasteful. And then remember that Paul Ehrlich had learned one lesson: change an apparently useless dye, a little, and it turns from a merely pretty color into *something* of a cure. That was enough to drive forward this too confident man.

All the time the laboratory was growing. To the good people of Frankfort Paul Ehrlich was a savant who understood all mysteries, who probed all the riddles of nature, who forgot everything. And how the people of Frankfort loved him for being so forgetful! It was said that this Herr Professor Doktor Ehrlich had to write himself postal cards several days ahead to remind himself of festive events in his family. "What a human being!" they said. "What a deep thinker!" said the cabbies who drove him every morning to his Institute. "That must be a genius!" said the grind-organ musicians whom he tipped heavily once a week to play dance music in the garden by the laboratory. "My best ideas come when I hear gay music like that," said Paul Ehrlich, who detested all highbrow music and literature and art. "What a democratic man, seeing how great he is!" said the good people of Frankfort, and they named a street after him. Before he was old he was legendary!

Then the rich people worshiped him. A great stroke of luck came in 1906. Mrs. Franziska Speyer, the widow of the rich banker, Georg Speyer, gave him a great sum of money to build the Georg Speyer House, to buy glassware and mice and expert chemists, who could put together the most complicated of his darling dyes with a twist of the wrist, who could make even the crazy drugs that Ehrlich invented on paper. Without this Mrs. Franziska Speyer, Paul Ehrlich might very well never have molded those magic bullets, for that was a job—you can watch what a job!—for a *factory* full of searchers. Here in this new Speyer House Ehrlich lorded it over chemists and microbe hunters like the president of a company that turned out a thousand automobiles a day. But he was really old-fashioned, and never pressed buttons. He was always popping into one or another of the laboratories every conceivable time of the day, scolding his slaves, patting them on the back, telling them of howling blunders he himself had made, laughing when he was told that his own assistants said he was crazy. He was everywhere! But there was always one way of tracking him down, for ever and again his voice could be heard, bawling down the corridors:

"Ka-de-reit! . . . Ci-gars!" or "Ka-de-reit! . . . Min-er-al wa-ter!"

5

The dyes were a great disappointment. The chemists muttered he was an idiot. But then, you must remember Paul Ehrlich read books. One day, sitting in the one chair in his office that wasn't piled high with them, peering through chemical journals like some Rosicrucian in search of the formula for the philosopher's stone, he came across a wicked drug. It was called "Atoxyl" which means: "Not poisonous." Not poisonous? Atoxyl had *almost* cured mice with sleeping sickness. Atoxyl had killed mice without sleeping sickness. Atoxyl had been tried on those poor darkies down in Africa. It had not cured them, but an altogether embarrassing number of those darkies had gone blind, stone blind, from Atoxyl before they had had time to die from sleeping sickness. So, you see, this Atoxyl was a sinister medicine that its inventors—had they been living—should have been ashamed of. It was made of a benzene ring, which is nothing more than six atoms of carbon chasing themselves round in a circle like a dog running round biting the end of his tail, and four atoms of hydrogen, and some ammonia and the oxide of arsenic—which everybody knows is poisonous.

"We will change it a little," said Paul Ehrlich, though he knew the chemists who had invented Atoxyl had said it was so built that it couldn't be changed without spoiling it. But every afternoon Ehrlich fussed around alone in his chemical laboratory, which was like no other chemical laboratory in the world. It had no retorts, no beakers, no flasks nor thermometers nor ovens—no, not even a balance! It was crude as the prescription counter of the country druggist (who also runs the postoffice) excepting that in its middle stood a huge table, with ranks and ranks of bottles—bottles with labels and bottles without, bottles with scrawled unreadable labels and bottles whose purple contents had slopped all over the labels. But that man's memory remembered what was in every one of those bottles! From the middle of this jungle of bottles a single Bunsen burner reared its head and spouted a blue flame. What chemist would not laugh at this laboratory?

Here Paul Ehrlich dabbled with Atoxyl, shouting: "Splendid!", growling: "Un-be-liev-a-ble!", dictating to the long-suffering Miss Marquardt, bawling for the indispensable Kadereit. In that laboratory, with a chemical cunning the gods sometimes bestow on searchers who could never be chemists, Paul Ehrlich found *that you can change Atoxyl*, not a little but a lot, that it can be built into heaven knows how many entirely unheard-of compounds of arsenic, without spoiling the combination of benzene and arsenic at all!

"I can change Atoxyl!" Without his hat or coat Ehrlich hurried out of this dingy room to the marvelous workshop of Bertheim, chief of his chemist slaves. "Atoxyl can be changed—maybe we can change it into a hundred, a thousand new compounds of arsenic!" he exclaimed. . . . "Now, my dear Bertheim," and he poured out a thousand fantastic schemes. Bertheim? He could not resist that "Now my dear Bertheim!"

For the next two years the whole staff, Japs and Germans, not to mention some Jews, men and white rats and white mice, not to mention Miss Marquardt and Miss Leupold— and don't forget Kadereit!—toiled together in that laboratory which was like a subterranean forge of imps and gnomes. They tried this, they did that, with six hundred and six—that is their exact number—different compounds of arsenic. Such was the power of the chief imp over them, that this staff never stopped to think of the absurdity and the impossibility of their job, which was this: to turn arsenic from a pet weapon of murderers into a cure which no one was sure could exist for a disease Ehrlich hadn't even dreamed might be cured. These slaves worked as only men can work when they are inspired by a wrinkle-browed fanatic with kind gray eyes.

They changed Atoxyl! They developed marvelous compounds of arsenic which—hurrah!—would really cure mice. "We have it!" the staff would be ready to shout, but then, worse luck, when the fell trypanosomes of the mal de Caderas had gone, those marvelous cures turned the blood of the cured mice to water, or killed them with a fatal jaundice. . . . And— who would believe it?—some of those arsenic remedies made mice dance, not for a minute but for the rest of their lives round and round they whirled, up and down they jumped. Satan himself could not have schemed a worse torture for creatures just saved from death. It seemed ridiculous, hopeless, to try to find a perfect cure. But Paul Ehrlich? He wrote:

"It is very interesting that the only damage to the mice is that they become dancing mice. Those who visit my laboratory must be impressed by the great number of dancing mice it entertains. . . ." He was a sanguine man!

They invented countless compounds, and it was a business for despair. There was that strange affair of the arsenic fastness. When Ehrlich found that one big dose of a compound was too dangerous for his beasts, he tried to cure them by giving them a lot of little doses. But, curse it! The trypanosomes became *immune* to the arsenic, and refused to be killed off at all, and the mice died in droves. . . .

Such was the grim procession through the first five hundred and ninety-one compounds of arsenic. Paul Ehrlich kept cheering himself by telling himself fairy stories of marvelous new cures, stories that God and all nature could prove were lies. He drew absurd diagrams for Bertheim and the staff, pictures of imaginary arsenical remedies that they in their expert wisdom knew it was impossible to make. Everywhere he made pictures for his boys—who knew more than he did—on innumerable reams of paper, on the menu cards of restaurants and on picture post cards in beer halls. His men were aghast at his neglect of the impossible; they were encouraged by his indomitable mulishness. They said: "He is so enthusiastic!" and became enthusiastic with him. So, burning his candle at both ends, Paul Ehrlich came, in 1909, to his day of days.

6

Burning his candle at both ends, for he was past fifty and his time was short, Paul Ehrlich stumbled onto the famous preparation 606—though you understand he could never have found it without the aid of that expert, Bertheim. Product of the most subtle chemical synthesis was this 606, dangerous to make because of the peril of explosions and fire from those constantly present ether vapors, and so hard to keep—the least trace of air changed it from a mild stuff to a terrible poison.

That was the celebrated preparation 606, and it rejoiced in the name: "Dioxy-diamino-arsenobenzol-dihydro-chloride." Its deadly effect on trypanosomes was as great as its name was long. At a swoop one shot of it cleaned those fell trypanosomes of the mal de Caderas out of the blood of a mouse—a wee bit of it cleaned them out without leaving a single one to carry news or tell the story. And it was safe! So safe—though it was heavily charged with arsenic, that pet poison of murderers. It never made mice blind, it never turned their blood to water, they never danced—it was safe!

"Those were the days!" muttered old Kadereit, long after. Already in those days he was growing stiff, but how he stumped about taking care of the "Father." "*Those* were the days, when we discovered the 606!" And they were the days —for what more hectic days (always excepting the days of Pasteur) in the whole history of microbe hunting? 606 was safe, 606 would cure the mal de Caderas, which was nice for mice and the hindquarters of horses, but what next? Next was that Paul Ehrlich made a lucky stab, that came from reading a theory with no truth in it. First Paul Ehrlich read—it had happened in 1906—of the discovery by the German zoölogist, Schaudinn, of a thin pale spiral-shaped microbe that looked like a corkscrew without a handle. (It was a fine discovery and Fritz Schaudinn was a fantastic fellow, who drank and saw weird visions. I wish I could tell you more of him.) Schaudinn spied out this pale microbe looking like a corkscrew without a handle. He named it the *Spirocheta pallida*. He proved that this was the cause of the disease of the loathsome name.

Of course Paul Ehrlich (who knew everything) read about that, but it particularly stuck in Ehrlich's memory that Schaudinn had said: "This pale spirochete belongs to the animal kingdom, it is not like the bacteria. Indeed, it is closely related to the trypanosomes. . . . Spirochetes may sometimes turn into trypanosomes. . . ."

Now, it was hardly more than a guess of that romantic Schaudinn that spirochetes had anything to do with trypanosomes, but it set Paul Ehrlich aflame.

"If the pale spirochete is a cousin of the trypanosome of the mal de Caderas—then 606 ought to hit that spirochete. . . . What kills trypanosomes should kill their cousins!" Paul Ehrlich was not bothered by the fact that there was no proof these two microbes were cousins. . . . Not he. So he marched towards his day of days.

He gave vast orders. He smoked more strong cigars each day. Presently regiments of fine male rabbits trooped into the Georg Speyer House in Frankfort-on-the-Main, and with these creatures came a small and most diligent Japanese microbe hunter, S. Hata. This S. Hata was accurate. He was capable. He could stand the strain of doing the same experiment a dozen times over and he could, so nimble was this S. Hata, do a dozen experiments at the same time. So he suited the uses of Ehrlich, who was a thorough man, do not forget it!

Hata started out by doing long tests with 606 on spirochetes not so pale or so dangerous. There was that spirochete fatal to chickens. . . . The results? "Un-heard . . . of! In-cred-i-ble!" shouted Paul Ehrlich. Chickens and roosters whose blood swarmed with that microbe received their shot of 606. Next day the chickens were clucking and roosters strutting— it was superb. But that disease of the loathsome name?

On the 31st of August, 1909, Paul Ehrlich and Hata stood before a cage in which sat an excellent buck rabbit. Flourishing in every way was this rabbit, excepting for the tender skin of his scrotum, which was disfigured with two terrible ulcers, each bigger than a twenty-five-cent piece. These sores were caused by the gnawing of the pale spirochete of the disease that is the reward of sin. They had been put under the skin of that rabbit by S. Hata a month before. Under the microscope—it was a special one built for spying just such a thin rogue as that pale microbe—under this lens Hata put a wee drop of the fluid from these ugly sores. Against the blackness of the dark field of this special microscope, gleaming in a powerful beam of light that hit them sidewise, shooting backwards and forwards like ten thousand silver drills and augers, played myriads of these pale spirochetes. It was a pretty picture, to hold you there for hours, but it was sinister—for what living things can bring worse plague and sorrow to men?

Hata leaned aside. Paul Ehrlich looked down the shiny tube. Then he looked at Hata, and then at the rabbit.

"Make the injection," said Paul Ehrlich. And into the ear-vein of that rabbit went the clear yellow fluid of the solution of 606, for the first time to do battle with the disease of the loathsome name.

Next day there was not one of those spiral devils to be found in the scrotum of that rabbit. His ulcers? They were drying already! Good clean scabs were forming on them. In less than a month there was nothing to be seen but tiny scabs—it was like a cure of Bible times—no less! And a little while after that Paul Ehrlich could write:

"It is evident from these experiments that, if a large enough dose is given, the spirochetes can be destroyed *absolutely and immediately with a single injection!*"

This was Paul Ehrlich's day of days. This was the magic bullet! And what a safe bullet! Of course there was no danger in it—look at all these cured rabbits! They had never turned a hair when Hata shot into their ear-veins doses of 606 three times as big as the amount that surely and promptly cured them. It was more marvelous than his dreams, which all searchers in Germany had smiled at. Now *he* would laugh! "It is safe!" shouted Paul Ehrlich, and you can guess what visions floated into that too confident man's imagination. "It is safe —perfectly safe!" he assured every one. But at night, sitting in the almost unbreathable fog of cigar smoke in his study, alone, among those piles of books and journals that heaped up fantastic shadows round him, sitting there before the pads of blue and green and yellow and orange note paper on which every night he scrawled hieroglyphic directions for the next day's work of his scientific slaves, Paul Ehrlich, noted as a man of action, whispered:

"Is it safe?"

Arsenic is the favorite poison of murderers. . . . "But how wonderfully we have changed it!" Paul Ehrlich protested.

What saves mice and rabbits might murder men. . . . "The step from the laboratory to the bedside is dangerous—but it must be taken!" answered Paul Ehrlich. You remember his gray eyes, that were so kind.

But, heigho! Here was the next morning, the brave light of the bright morning. Here was the laboratory with its cured rabbits, here was that wizard, Bertheim—how he had twisted that arsenic through all these six hundred and six compounds. That man could not go wrong. So many of them had been dangerous that this six hundred and sixth one *must* be safe. . . . Bravo! Here was the mixed good smell of a hundred experimental animals and a thousand chemicals. Here were all these men and women, how they believed in him! So, let's go! Let us try it!

At bottom Paul Ehrlich was a gambler, as who of the great line of the microbe hunters has not been?

And before that sore on the scrotum of the first rabbit had shed its last scab, Paul Ehrlich had written to his friend, Dr. Konrad Alt: "Will you be so good as to try this new preparation, 606, on human beings with syphilis?"

Of course Alt wrote back: "Certainly!" which any German doctor—for they are right hardy fellows—would have replied.

Came 1910, and that was Paul Ehrlich's year. One day, that year, he walked into the scientific congress at Koenigsberg, and there was applause. It was frantic, it was long, you would think they were never going to let Paul Ehrlich say his say. He told of how the magic bullet had been found at last. He told of the terror of the disease of the loathsome name, of those sad cases that went to horrible disfiguring death, or to what was worse—the idiot asylums. They went there in spite of mercury—mercury fed them and rubbed into them and shot into them until their teeth were like to drop out of their gums. He told of such cases given up to die. One shot of the compound six hundred and six, and they were up, they were on their feet. They gained thirty pounds. They were clean once more—their friends would associate with them again. . . . Paul Ehrlich told, that day, of healings that could only be called Biblical! Of a wretch, so dreadfully had the pale spirochetes gnawed at his throat that he had had to be fed liquid food through a tube for months. One shot of the 606, at two in the afternoon, and at supper time that man had eaten a sausage sandwich! There were poor women, innocent sufferers from the sins of their men—there was one woman with pains in her bones, such pains she had been given morphine every night for years, to give her a little sleep. One shot of compound six hundred and six. She had gone to sleep, quiet and deep, with no morphine, that very night. It was Biblical, no less. It was miraculous—no drug nor herb of the old women and priests and medicine men of the ages had ever done tricks like that. No serum nor vaccine of the modern microbe hunters could come near to the beneficent slaughterings of the magic bullet, compound six hundred and six.

Never was there such applause.

Never has it been better earned, for that day Paul Ehrlich —forget for a moment the false hopes raised and the troubles that followed—that day Paul Ehrlich had led searchers around a corner.

But, to every action there is an equal and opposite reaction. What is true in the realm of lifeless things is true in the lives of such men as Paul Ehrlich. The whole world bawled for Salvarsan. That was what Ehrlich—we must forgive him his grandiloquence—called compound six hundred and six. Then, in the laboratory of the Georg Speyer House, Bertheim and ten assistants—worn these fellows were before they started it —turned out hundreds of thousands of doses of this marvelous stuff. They did the job of a chemical factory in their small laboratory, in the dangerous fumes of ether, in the fear that one little slip might rob a hundred men and women of life, for it was two-edged stuff, that Salvarsan. And Ehrlich? Now he was only a shell of a man, with diabetes—and why did he keep on smoking more cigars?—now Ehrlich burned the candle in the middle.

He was everywhere in the Georg Speyer House. He directed the making of compounds that would be still more wonderful—so he hoped. He chased around so that even Kadereit couldn't keep track of him. He dictated hundreds of enthusiastic letters to Martha Marquardt, he read thousands of letters from every corner of the world, he kept records, careful records they were too, of every one of the sixty-five thousand doses of Salvarsan injected in the year 1910. He kept them— this was like that strangely systematic man!—on a big sheet of paper tacked to the inside of the cupboard door of his office, from the top to the bottom of that door in tiny scrawls, so that he had constantly to squat on his heels or stretch up on tiptoe and strain his eyes to read them.

As the list grew, there were records of most extraordinary cures, but there were reports it was not pleasant to read, too, records that told of hiccups and vomitings and stiffenings of legs and convulsions and death—every now and then a death in people who had no business dying, coming right after injections of the Salvarsan.

How he worked to explain them! How he wore himself to a shred to avoid them, for Paul Ehrlich was not a hard-boiled man. He made experiments; he conducted immense correspondences in which he asked minute questions of just how the injections had been made. He devised explanations, on the margins of the playing cards he used for his games of solitaire each evening, on the backs of those blood-and-thunder murder mysteries that were the one thing he read—so he imagined—to rest. But he never rested! Those disasters pursued him and marred his triumph. . .

The wrinkles deepened to ditches on his forehead. The circles darkened under those gray eyes that still, but not so often, danced with that owlish humor.

So this compound six hundred and six, saving its thousands from death, from insanity, from the ostracism worse than death that came to those sufferers whose bodies the pale spirochete gnawed until they were things for loathing, this 606 began killing its tens. Paul Ehrlich wore his too feeble body to a shadow, trying to explain a mystery too deep for explanation. There is no light on that mystery now, ten years after Ehrlich smoked the last of his black cigars. So it was that this triumph of Paul Ehrlich was at the same time the last disproof of his theories, which were so often wrong. "Compound six hundred and six unites chemically with the spirochetes and kills them—it does not unite chemically with the human body and so can do no damage!" That had been his theory. . . .

But alas! What is the chemistry of what this subtle 606 does to the still more subtle—and unknown—machine that is the human body? Nothing is known about it even now. Paul Ehrlich paid the penalty for his fault—which may be forgiven him seeing the blessings he has brought to men—his fault of not foreseeing that once in every so many thousands of bodies a magic bullet may shoot two ways. But then, the microbe hunters of the great line have always been gamblers: let us think of the good brave adventurer Paul Ehrlich was and the thousands he has saved.

Let us remember him, trail-breaker who turned a corner for microbe hunters and started them looking for magic bullets. Already (though it is too soon to tell the whole story) certain obscure searchers, some of them old slaves of Paul Ehrlich, sweating in the great dye factories of Elberfeld, have hit upon a most fantastical drug. Its chemistry is kept a secret. It is called "Bayer 205." It is a mild mysterious powder that cures the hitherto always fatal sleeping sickness of Rhodesia and Nyassaland. That was the ill, you remember, that the hard man, David Bruce, fought his last fight, in vain, to prevent. It does outlandish things to the cells and fluids of the human body—you would say they were fibs and fairy tales if you heard the queer things that drug can do! But what is best, it slaughters microbes! It kills them beautifully, precisely, with a completeness that must make Paul Ehrlich wriggle in his grave—and when it doesn't kill microbes it *tames* them.

It is as sure as the sun following the dawn of to-morrow that there will be other microbe hunters to mold other magic bullets, surer, safer, bullets to wipe out for always the most malignant microbes of which this history has told. Let us remember Paul Ehrlich, who broke this trail. . . .

This plain history would not be complete if I were not to make a confession, and that is this: that I love these microbe hunters, from old Antony Leeuwenhoek to Paul Ehrlich. Not especially for the discoveries they have made nor for the boons they have brought mankind. No. I love them for the men they are. I say they *are*, for in my memory every man jack of them lives and will survive until this brain must stop remembering.

So I love Paul Ehrlich—he was a gay man who carried his medals about with him all mixed up in a box never knowing which ones to wear on what night. He was an impulsive man who has, on occasion, run out of his bedroom in his shirt tail to greet a fellow microbe hunter who came to call him out for an evening of wassail.

And he was an owlish man! "You say a great work of the mind, a wonderful scientific achievement?" he repeated after a worshiper who told him that was what the discovery of 606 was.

"My dear colleague," said Paul Ehrlich, "for seven years of misfortune I had one moment of good luck!"

Index